Freud and Psychoanalysis

Freud and Psychoanalysis

Six Introductory Lectures

JOHN FORRESTER

Edited by Lisa Appignanesi
Foreword by Darian Leader

polity

First published in 2023 by Polity Press

Polity Press
65 Bridge Street
Cambridge CB2 1UR, UK

Polity Press
111 River Street
Hoboken, NJ 07030, USA

ISBN-13: 978-1-5095-5811-7
ISBN-13: 978-1-5095-5812-4(pb)

A catalogue record for this book is available from the British Library.

Library of Congress Control Number: 2022948656

Typeset in 11 on 14pt Warnock Pro
by Cheshire Typesetting Ltd, Cuddington, Cheshire
Printed and bound in Great Britain by TJ Books Ltd, Padstow, Cornwall

The publisher has used its best endeavours to ensure that the URLs for external
websites referred to in this book are correct and active at the time of going to press.
However, the publisher has no responsibility for the websites and can make no
guarantee that a site will remain live or that the content is or will remain appropriate.

'In Memory of Sigmund Freud,' copyright 1940 and © renewed 1968 by W. H. Auden;
from Collected Poems by W. H. Auden, edited by Edward Mendelson. Used by
permission of Curtis Brown, Ltd. and Random House, an imprint and division of
Penguin Random House LLC. All rights reserved.

Every effort has been made to trace all copyright holders, but if any have been
overlooked the publisher will be pleased to include any necessary credits in any
subsequent reprint or edition.

For further information on Polity, visit our website:
politybooks.com

Contents

Editor's Preface

John Forrester knew his Freud backwards. That not only included the twenty-four volumes of the Standard Edition in English, the German originals and sometimes the French translations. It also included the voluminous correspondence that Freud engaged in with any number of friends and fellow practitioners, commentaries on Freud and Freudian history in various languages, the history of psychoanalysis and its migrations around the world. He followed the thinking that Freud provoked, from the work of Jacques Lacan, to Donald Winnicott and Wilfred Bion, to French philosophers such as Paul Ricœur and Jacques Derrida, to literary theory, cultural studies and feminism.

But unusually for a scholar of Freud and psychoanalysis, Forrester was also a historian and philosopher of science. He had started his studies as a laboratory chemist – not that he stayed in the laboratory for very long (though the interest did last long enough to show traces in his first graduate papers for the philosopher of science, Thomas Kuhn, at Princeton). At the University of Cambridge's renowned History and Philosophy of Science Department, where he spent most of his academic life, and which he headed from 2007 to 2013, Forrester was immersed in the various cross-over fields the subject covered, from the history of medicine and psychiatry and biomedical sciences to the emerging sciences, to the philosophies that accompany the ways we understand them, from Descartes to Wittgenstein to Austin, to name but a few signposts.

It is this last emphasis that makes Forrester's own thinking about Freud and psychoanalysis rare, as his books attest. Perhaps it also allowed him to see how Freud's own insistence on his role as a scientist, so much disputed, had more than a grain of historical truth in it.

Forrester was, dare I say, brilliant, often witty, and he certainly liked to provoke. I wish I could add to these many and luminous talents an ability reliably to switch on a tape recorder. Though he delivered many lectures over many decades, to students, colleagues and the general public, the introductory lectures reproduced in this book, on Freud and psychoanalysis, were, bar one other, the only ones recorded. On occasion, they also failed to fully record.

What follows in this collection are the reconstructions of the lectures Forrester offered annually to his undergraduate students. Forrester first taught Freud in 1974. For the next decade, in his teaching, he sought to introduce a Freud that could be as at home within the history of science as in literature. This meant approaching Freud in an unusually interdisciplinary way. He taught Freud and social theory, and psychoanalysis and femininity to students in the social and political sciences; Freud and Lacan in the faculties of English and modern languages; and Freud and the history of psychiatry in the Department of Experimental Psychology, and Natural Sciences Tripos. After taking up a permanent position in the Department of History and Philosophy of Science, Forrester continued teaching a wide range of classes: on Darwin and Darwinism; the history of thermodynamics; Bachelard and Foucault; gender and science; the history of psychiatry; science and the history of the world; the history of the human sciences;

heat, chance and psychoanalysis – to name but a few. He is probably the only person to have ever lectured on writing and the unconscious to students of literature and on the history of modern sexual science to students of anatomy. Suffice to say, Forrester's lectures were vast in their range and erudite in their coverage. It is this interdisciplinary nature of Forrester's account of Freud and psychoanalysis that makes what follows so illuminating and so distinctive as an introductory text.

The lectures collected here were delivered in Michaelmas Term 2012. Forrester had a habit of lecturing from bullet points (and, later, power point slides). He wrote no scripts. And though notes survive – and can be consulted in Forrester's papers (held at the Manuscript Division in the Albert Sloman Library at the University of Essex) – they cannot be fully reconstructed. Only these lectures can be posthumously published.

A brief further note on the text. It was our daughter, Katrina Forrester, who urged me to put her father's lectures into publishable shape. I was greatly helped in preparing what follows by Tasha Pick, who, with great flair, undertook the initial labour of transcription, which included the first hurdles of turning speech into prose. I took the task over from there, editing, adding material from notes, sometimes being driven to juggle, insert or write a few lines to carry the sense. Katrina then read and edited a draft of the full manuscript, checking against the recordings to ensure that it was faithful to the original spoken lectures, and helping to reconstruct the meaning of John's more elliptical notes and claims. Her help has been invaluable. John and I had worked, even written a book, together before, so I trust I haven't traduced his sense, or not too much. He would have smiled happily at the thought of his daughter's help and I imagine had a moment of proud wonder that all those dinner table arguments could have contributed to this volume. He would also have been extremely grateful, as am I, to Dany Nobus who helped with Lacan, and to Darian

Leader for his spirited Foreword, which so well captures John's intellectual style.

In what follows, much has been preserved from the original spoken lectures – tangents, free associations and all. Where there is a significant departure from the recording or his notes, it is mentioned in the endnotes.

I hope that what follows is a pleasure to read and also proves thought-provoking to anyone interested in Freud and psychoanalysis and their role in shaping our world. Freud's great invention – that talking and listening technology of two – plays its part in our times. It's as well to understand a little more about it – which is what the following pages help us to do.

Lisa Appignanesi

Foreword

A survey of Freud's work and its impact on twentieth-century thought ought to be the subject of a multivolume compendium, complete with thousands of pages and a list of references like a telephone directory. Yet here is a short book, made up of six lectures for undergraduates, that manages not only to provide an effortless introduction to Freud and his impact, but also to raise key questions about the methods, aims and status of psychoanalysis as both a scientific endeavour and a clinical practice.

John Forrester's grasp of Freud was celebrated. He had authored a pioneering study on the place of language in the origins of psychoanalysis, followed by a series of books and articles on the history of psychoanalysis and the role of speech and truth games in analytic practice and in the world of psychoanalysts. These were unusual works, bringing together the eye of the historian of science with a curiosity about gossip and bluffing, and, more generally, the sociology of the analytic world.

The scope of the lectures reflects these interests: Freud's theories are linked to the conceptual and ideological backgrounds of his time, as well as to biographical and cultural questions. Darwin and Meynert are here, but also Auden, Hitchcock and D. M. Thomas. Forrester shows not only how Freud helped shape twentieth-century views of infancy and childhood, sexuality and femininity, but also how some of his ideas were contested or reformulated by later thinkers. To do all this in such a slim and agile volume is quite an achievement.

Forrester returns several times in these lectures to the work of the anthropologist Ernest Gellner, *The Psychoanalytic Movement*, which explores the links between psychoanalysis and religion as belief systems. I remember attending the seminars at which Gellner first elaborated these ideas, with Forrester interrupting frequently to challenge or nuance Gellner's arguments. To the calculation that most of the world's population could have been analysed by 1980, assuming that a personal analysis could produce a practitioner of psychoanalysis who would then go on to analyse others, Forrester asked Gellner if he had any evidence that this was not in fact the case. As he gently and humorously pursued this dialogue, he showed how generalizations and 'facts' had to be approached critically, and how knowledge claims were always open to question.

This dialectical approach to learning shared something with that of Thomas Kuhn, with whom Forrester had studied as a graduate student. When Kuhn was stuck in his own doctoral dissertation on Copernican physics, he was in analysis, and explained later that it was the analytic work that allowed him to make his breakthrough: rather than searching for the coherence of the Copernican system, he realized that he should be looking for contradictions and inconsistencies in this system. Like the symptoms and slips that appear in analysis, these were the fault lines that would reveal what the underlying questions really were for Copernicus.

This was certainly Forrester's style, evident in these lectures, his writings and his conversations. He would push and probe a received opinion to see where it led, bringing in new data and insights to explore what in a belief system was only belief and what, in contrast, was perhaps something else. And this had a strange and rather unusual effect: in contrast to just about every other Freud scholar on the planet, it was never clear whether he was 'for' or 'against'.

Although such polarizations are ultimately unhelpful, and tend to generate reductionisms, people who devote their lives to the study of psychoanalysis are almost always either believers or iconoclasts. This becomes clear after a few pages of text or a few minutes of conversation, but this was never the case with Forrester. His passion and enthusiasm for Freud scholarship and the study of psychoanalysis was resolute and intractable, and yet he maintained right until the end less a critical distance than what we should term a critical engagement. Everyone who knew him could learn from this, and not a little of this style is present in these lectures.

Lacan famously opened his 'Écrits' by citing Buffon: 'Le style, c'est l'homme'. But Freud had corrected this many years previously, when he quipped 'Le style, c'est l'histoire de l'homme'. Just as psychoanalysis has taught us never to neglect human history and the vicissitudes of an individual life, so Forrester shows right through his work the importance of the past and how, for the Freudian century, it has played a part in shaping the future.

Darian Leader

Note on Text

All quotations from Freud, letters apart, come from *The Standard Edition of the Complete Psychological Works of Sigmund Freud*, 24 vols, ed. James Strachey in collaboration with Anna Freud, assisted by Alix Strachey and Alan Tyson (London: Hogarth Press and the Institute of Psycho-analysis, 1953–74). This is abbreviated after quotations to SE plus volume and page number.

Photograph showing Sigmund Freud recording his voice for the
BBC, 7 December, 1938. © Freud Museum London

Lecture 1

A Whole Climate of Opinion

Introducing Freud

I want to begin this lecture series with a much-quoted passage from W. H. Auden's poem, composed on hearing of Freud's death in 1939:

> If often he was wrong and, at times, absurd,
> to us he is no more a person
> now but a whole climate of opinion
>
> under whom we conduct our different lives:
> Like weather he can only hinder or help[1]

This passage brings into focus the idea that we inhabit a Freudian universe. Freud is the very climate in which we conduct our lives. He is pervasive in our culture. We cannot imagine an alternative way of seeing the world – even or perhaps especially when the Freudian origin of our views is lost or obscured. His work saturates our ways of seeing and thinking. In this sense, Freud can be compared with Copernicus or Darwin. The analogy is one Freud himself made in talking about the revolutions which had shaken man's sense of his own central place in the universe. Though Freud's status amongst these influential figures is contested, psychoanalysis, with its insistence on an unconscious life of which reason is only sporadically aware, did aim a bitter blow at our sense of being 'masters' in our own house. For Freud, the rational subject is at the mercy both of an unpredictable unconscious and the occluded memories from earliest life that shape him. Freud put the point in the following way in his Introductory Lectures:

> Human megalomania will have suffered its third and most wounding blow from the psychological research of the present time which seeks to prove to the ego that it is not even master in its own house, but must content itself with scanty information of what is going on unconsciously in its mind. We psycho-analysts were not the first and not the only ones to utter this call to introspection; but it seems to be our fate to give it its most forcible expression and to support it with empirical material which affects every individual. (SE XVI: 284)

How did Freud conceive of his own project? On 7 December 1938, just nine months before his death, the BBC recorded Freud in his new home in London:

> *Im Alter von 82 Jahren verließ ich in Folge der deutschen Invasion mein Heim in Wien und kam nach England, wo ich mein Leben in Freiheit zu enden hoffe* ... My name is Sigmund Freud.

> [At the age of 82, I fled my home in Vienna as a result of the German invasion and came to England, where I hope to end my life in freedom.]

The rest of the archive clip of the speaking Freud is in his remarkably good English, undoubtedly honed in listening to a substantial number of English and American patients over his practising lifetime:

> I started my professional activity as a neurologist trying to bring relief to my neurotic patients. Under the influence of an older friend and by my own efforts, I discovered some important new facts about the unconscious in psychic life, the role of instinctual urges and so on. Out of these findings grew a new science, psychoanalysis, a part of psychology and a new method of treatment of the neuroses. I had to pay heavily for this bit of good luck. People did not believe in my facts and thought my theories unsavoury. Resistance was strong and unrelenting. In the end, I succeeded in acquiring pupils and building up an International Psychoanalytic Association. But the struggle is not yet over.[2]

Freud's combative and ambitious tone is unmistakable here. The word 'unsavoury' is perhaps particularly revealing and well chosen in the context: it implies a sensuous, bodily relationship to theory. This is something Freud emphasized throughout his work.

How might we understand Freud today? The vast scope of his work and interests is evident in the numerous epithets that might be used to describe him: scientist, doctor, inventor, cultural critic, writer, moral exemplar, scientific entrepreneur. Freud thought of himself straightforwardly as a scientist. 'What else could I be?', he asked. His first ambitions lay in research, but needs demanded a profession that would allow him to subsidize his parental family, then marry and make his own. He qualified as a doctor in 1881 with a medical degree from the University of Vienna. In 1886, he opened his first consulting-room practice, moving to the now famous Berggasse 19 in 1891, where he worked and lived with his family for the next forty-seven years before fleeing to London. There, at Maresfield Gardens, he continued to see patients until death overtook him in September 1939. His invention of the 'talking cure', with his earliest mentor Josef Breuer and at the inspiration of the first patients, the 'hysterics' he called his teachers, forms the core of what would become the psychotherapeutic industry.

Freud was also a cultural and social critic of some importance, particularly in later works such as *Civilization and Its Discontents* (1930) – an essay on the inevitable conflict between individual desires and the restrictions civilization imposes – which remains a key text in courses throughout the Western world. Freud never received a Nobel Prize for either science or literature – unlike his contemporaries Bertrand Russell and Winston Churchill – a fact perhaps surprising given the significant impact of his work. He was however a wonderful writer, is still a bestseller amongst the classics, and did receive the Goethe Prize for Literature in 1930.[3]

Disagreements about Freud's moral character as well as his reputation have long been in play. By some, he is seen as a moral exemplar – a good doctor: steadfast, stoical, humorous, in relation to his patients and in his personal life. His private correspondence, as well as his psychoanalytic writings and the memoirs of patients, give much evidence of this. As an approach to the legacy of his work, however, the emphasis on his moral exemplarity has often served as an invitation to his critics. These characterize Freud as a liar, a fraud and a cheat, and invoke such attributed failings to reject the whole of psychoanalysis. In this way, an attack on the character of the founder of psychoanalysis becomes an attack on the whole edifice of psychoanalysis, its theory and practice. An illustrative parallel can be drawn, perhaps, with the American physicist Robert Millikan, whose experiments to determine the electric charge of the electron were later partly discredited due to fraud allegations.[4] If Millikan were treated in the same way that Freud is, the fact that Millikan fudged some of his data would enable critics to argue that the electron itself did not exist. These are the kind of arguments that are made beginning from

Freud's moral exemplarity: yet if he fudged or truncated a case history, does it really follow that the unconscious or repression do not exist?

As for Freud's status as scientific entrepreneur, that is undeniable. Freud built up an International Psychoanalytic Association with many thousands of members around the world, creating in the process a market for therapy. But if Freud was an entrepreneurial inventor, what kind of invention was psychoanalysis? It was one that grew organically out of Freud's medical practice. At the beginning of his career, Freud wrote for his colleagues. He had a medical readership in mind. Though medicine itself was not a science, it intersected with many sciences and neurology, and with the disciplines of psychiatry and psychology, both of which were going through complicated and important periods of growth in the late nineteenth century. Freud's psychoanalysis represented a contribution to the development of psychology. But his work also fell under the field of psychiatry, though in a complex way. The term 'psychiatry' was first used in German in the early nineteenth century and came to have a very specific set of meanings by its close. Freud's psychiatric work consisted both in working with patients during a two-year stint as a hospital doctor, and as a leading scientific researcher. Put very bluntly, psychiatry, for Freud, means brains, not minds; it also means the kind of diagnostic classification used in large-scale confining institutions. Freud's medical practice initially intersected with the brain sciences so in vogue in his day and in our own. But though he loved his work as a brain scientist, Freud ended up following a different route: within medicine, he became a specialist in mental and nervous disorders, as they were then described in English.

Introducing psychoanalysis

As psychoanalysis grew, particularly from 1900 onwards, it developed relationships to wider elite scientific and social movements. I mention the social and the scientific in the same breath to point out that psychoanalysis was very quickly of interest to artistic and cultural figures in Vienna and elsewhere. Psychoanalysis became closely connected not only with new scientific developments, but with cultural, even political, revolutions. It broadened out to encompass a complicated set of ideas and practices, becoming a movement that would even be described as a secular religion. This is the formula put forward by Ernest Gellner, an outstanding anthropologist, in his book *The Psychoanalytic Movement* (1985). Gellner proposes that psychoanalysis performs a similar function to religion, but without its trappings, since it is essentially scientific. This prompts the question: under what circumstances does science become a religion? Is psychoanalysis a good example of this? Does it teach us something about how science becomes religion? Or is there a necessary immiscibility between science and religion, like oil and water, existing in parallel worlds but unable to mix?

At the core of psychoanalysis are two crucial concepts: therapy and sex, or what Freud called libido. I want to give a sense of how this might work at the most basic of levels today with an anecdote from a clinical practitioner whom I've known for a very long time – a medically qualified psychiatrist and practising psychoanalyst who was working in an NHS hospital at the time of this story. As an analyst, he was treated as an eccentric outsider, the lowest of the low in terms of the hospital's pecking order. He was referred a case of a severely depressed farmer who had been treated with all the modern methods: electroconvulsive therapy (ECT), which was then meant to be the best treatment for endogenous depression; antidepressants; and cognitive behavioural therapy (CBT), which is what the NHS now routinely prescribes as therapy.

The patient was finally turned over to the psychoanalyst as a form of baiting the enemy.

The analyst established that the patient's depression had come on a short time after the death of his wife following a long illness. When she was dying, his wife had said: 'You must never go with another woman.' Some weeks after her death, the farmer began to experience disturbed sleep, including sleepwalking. He told his analyst that during one of his sleepwalking episodes, he'd woken up to find himself in the barn having sexual intercourse with a pig. At this point, the analyst said: 'You reacted to your wife's death with an inner determination to fulfil her dying wishes. But you don't have to.' The effect of this intervention on the patient was striking and undeniable, much to the amazement of the orthodox psychiatrists, who had been oblivious to this story because they don't ask patients those kinds of questions.

What does this tell us? Perhaps the first thing to say is that although this is a psychoanalytic case of a very small order, it expresses the fact that psychoanalytic cases are often too amazing to be true. They are beautiful embodiments of the notion that fact is stranger than fiction. This story obviously has no evidential standing whatsoever in scientific terms. It's just a story. In fact, it might even be fiction. Who knows? How do I know that the psychoanalyst who told me this story didn't make it up? It certainly doesn't conform to canons of clinical trials and randomized controls. But what it does is to make absolutely clear the motivations and the mechanisms by which an obscure illness is generated. It is predicated on the assumption that depression has a meaning; it is an inner act on the part of the patient. Where is the evidence for this, the proof? What counts as proof here is a recourse to ordinary psychological explanations, to what is sometimes called by philosophers of mind 'folk psychology'.

The case described above also conforms perfectly to Freud's definition of the process of melancholy, his term for depression in its relation to mourning, as set out in 'Mourning and melancholia' (1917). For Freud, depression in mourning occurs through an attack that has been displaced onto the ego rather than centring on the lost object, here the wife. Depression is an attack on myself because I do not want to attack my prized object, external to the self – someone I can no longer love or who no longer loves me, or some situation of deep importance in my erotic and general life, which gave meaning to the world but is now lost. In this case, the farmer, it seems, fell ill of his fidelity to his wife. In order to remain faithful to his wife's words, he had to deny his own sexual needs. Importantly, he had not 'gone with another woman' and had remained faithful to his wife by not straying, by at least staying within his own farm. The psychoanalyst's intervention was to say, you can separate off the question of fidelity to your wife from your living being as a sexual man.

Psychoanalysis exists not only as a therapeutic practice – one that is accepted or rejected by the medical profession – but as a player with its own independent life in the wider cultural sphere. A good example of this lies in the literally thousands of *New Yorker* cartoons about psychoanalysis. One of my favourites is particularly resonant for a historian of science because it sums up the three paths for psychology since the late nineteenth century. A well-dressed, bespectacled ape sits in the analyst's chair taking notes on the brain-in-a-jar lying on the analytic couch, the couch being one of the trademark 'tools' of the profession. The ape alludes to the importance of evolutionary accounts of animal behaviour to the development of psychology. The brain sits in a jar, ready to be sliced open by the neuroscientists. One can imagine a scanner – a Jules Verne version of a scanner, perhaps – imaging it. And then you have the couch on which the patient relines, a position that facilitates the particular kind of free associative speech that became the basis for the psychoanalytic method. The ape, finally, is Freud – well, not quite, but he's certainly listening to what the brain has to tell him. All three models of psychology are present here: brain science, evolutionary science and 'talk' science.

Such cartoons are also examples of how, through the course of the twentieth century, psychoanalysis entered popular culture, generating, along the way, not only *New Yorker* cartoons, but also hundreds of films, novels and advertising. Adam Curtis's documentary series *The Century of the Self* (2002) presents a mass of such examples to conjure a wonderfully paranoid vision of twentieth-century culture, making much of the fact that one of the founders of advertising in the United States was Freud's nephew, Edward Bernays. Curtis traces the close affiliation between psychoanalysis and such 'hidden persuaders': adverts from as early as the 1930s often allude to or use varieties of Freudian theory. The language of psychoanalysis quickly infiltrated everyday life and discourse, particularly in the English-speaking world. Freud's use of the story of Oedipus and his 'smutty' interpretation of its relation to family life became near ubiquitous in the fifties, the subject of jokes as well as t-shirts. Terms such as 'Freudian slip' have been standard phrases since the 1930s.

Psychoanalysis, then, has diverse and multiple cultural presences. It absorbs film, literature, science, politics. If in Vienna and Berlin in the 1920s psychoanalysis spread into schools, polyclinics and medico-sexual outreach programmes as various forms of psychotherapy, by the second half of the twentieth century – and particularly, to start with, in the United States – the spread of the Freudian lexicon goes hand in hand with what has become known as 'therapy culture'. Therapy culture begins with psychotherapy but disseminates much more widely with the growth of the health and well-being industries that built gyms, proliferated heath trends and new practices (like jogging, which I date to the 1970s, though the word was introduced in 1964). Therapy culture also infiltrated other formal and informal institutions, like prisons and schools, and reached into social work, social media and even public policy.

In this way, psychoanalysis is distinct from many sciences. But like any science, it is first and foremost a body of knowledge bound together with certain practices. It is also a profession; individuals who join that profession conform to its standards, set up institutions associated with it and garner cultural capital in the process. This is true, clearly, for all scientific and medical institutions. Yet psychoanalysis also has a very different set of cultural influences, going far beyond its formal professional practitioners.

People are often perplexed by what psychoanalysis is for this reason. It goes beyond one set of institutions or one professional practice associated with elite practitioners. It doesn't conform to the standard models of how science should behave. One could describe it as a kind of guerrilla science, or, as Gellner suggested, a new religion – or perhaps the form religion takes in a secular and scientific culture. In his *Introductory Lectures*, Freud wryly draws religion in to note that psychoanalysis cannot hope to compete with Lourdes, the French town famous for its miracle faith cures of fatal diseases, where Catholics have gathered to be healed since the late nineteenth century. It is true that Lourdes successes are astonishing: it really does seem to cure people of cancer and other maladies. Psychoanalysis may be playing in the same field, trying to harness psychological forces to effect extraordinary cures of both mental and physical illness. But Freud says that psychoanalysis can never compete with Lourdes. This thought raises the question of whether this may be true for biomedical science in general. If the biomedical sciences are somehow trying to rival Lourdes and find as effective a placebo, the road may prove long indeed.

There is a last point to make here briefly about the cultural status of psychoanalysis in the context of secularization. One narrative of modernity views religion losing its cultural hold and foundations in the late nineteenth century and being replaced by a hedonistic and aesthetic culture, with the subsequent ascendance of a cultural form according to which the meaning of life is beauty and pleasure. On this narrative, one might see psychoanalysis as the last phase of the European Enlightenment – the Enlightenment turned inwards. The story of psychoanalysis becomes part of the story of the production of culture in the 'West' after religion loses its foundations.

Psychoanalysis and science

There is another way of framing the history of psychoanalysis. Psychology develops as a discipline in three important modes during Freud's lifetime. E. G. Boring, in his extensive *A History of Experimental Psychology* (1929), implicitly makes the claim that the only branch of psychology that is truly 'scientific' is experimental psychology, the work that takes place in laboratories and universities.[5] This is typical in traditional accounts of the discipline. In an excellent historical work, *Constructing the Subject* (1990), Kurt Danziger turns away from these conventional assumptions, proposing instead that even if you agree that scientific psychology comes into existence in the middle of the nineteenth century, there are multiple programmes involved in its development.[6]

First, there is the academic programme of experimental psychology with its close ties to experimental physiology in the German universities from the 1850s onwards. Wilhelm Wundt is the founding father here. He set up the first laboratory in experimental psychology at the University of Leipzig in 1879. Wundt's work was based on establishing criteria of objective measurement: through experimental observation, subjective meaning was to be eliminated. But this story is immediately complicated by the problem of how to experiment in psychology. What is it that is being measured? The experimenter, after all, does something to the subject of the experiment. Let's call this subject Nameless. The experimenter, often using a word or a signal, measures how Nameless responds. In this sense, the experiment measures through the experimenter how Nameless monitors him/herself and their reactions. Inherent to experimental psychology is a set-up that relies on a basis of training in expert self-monitoring. If you're sticking two needles into someone's back and trying to measure how far apart they have to be in order for the subject to be able to distinguish between the two, the subject has to become an expert on reporting sensation in the back. Psychological research is predicated on this idea: it involves training subjects to become experts. The most effective way to observe this dynamic is to repeat the experiment with the researcher and the subject having swapped seats. The two positions are completely asymmetrical. Accurately reporting one's responses is as learnt a behaviour as setting up the experiment and asking for or documenting the response. The study of psychology requires this kind of introspection, plus laboratory observation. In the late nineteenth century, psychology and 'introspectionism' are seen as one and the same.

The second psychology programme to develop in this period is Galtonian psychology, named after the English scientist Francis Galton, a cousin of Darwin's. This is a psychology of large numbers, one that surveys the characteristics and features of populations. It takes off from the development of statistics in the nineteenth century. The famous starting point for this field of study is the measuring of the height of recruits in the Scottish army. Instrumental to this measurement was the newly discovered Gaussian curve, named after the German mathematician, Karl Gauss. The foundation of this programme was to show the bell curve distribution of the properties of populations. Galton was interested in a psychology not of the individual, but of the mass. His psychology was located not in the university, but in large institutions, in factories and schools.

It is closely linked to the social developments of the period. The first law requiring children to attend school until the age of 11 was passed in 1870. The Galtonian project is about these new populations and their properties, the most famous being the IQ test, which very controversially sought to measure the distribution of 'intelligence' across genders, races and so on.

The third kind of psychology that Danziger points to is that of clinical investigation. This is closely related to medicine. Here the influential French neurologist Jean-Martin Charcot, Freud's teacher in Paris, and Freud himself both play star parts. This is medical psychology – the term used in Britain until the 1920s for what is later called psychiatry. It is fundamentally based on individual cases. It differs from experimental psychology in its explanatory frameworks, which are largely statistical. Medical psychology is interested in isolating specific causal patterns, the particular origins and effects of symptoms. It draws on a medical model of causal inference that can be seen, for instance, in the developments in bacteriology in the 1880s, which established the idea that the necessary and sufficient condition for typhoid was a particular bacterium. This model is transposed onto explorations of the brain in medical psychology.

Psychoanalysis emerges as a new psychology in the early twentieth century. 'New psychology' is something of a slippery term, with different usages in Britain and the US. In England, it's used from around 1915 to refer to Freud, Jung and some others. In America, it is used in the 1880s to describe experimental psychology. I am using it here in the English sense. At its most fundamental level, psychoanalysis emerges as a new psychology interested in the instinctual and the irrational. Freudian theory could essentially be described as a rational understanding of the irrational, or a rational encounter with the force of desire. Later we see the idea of 'depth psychology' or *Tiefenpsychologie*, to use the German term. Two important features of Freud's work are at play in this notion of 'depth':

chronological depth, the past's relationship to the present and, relatedly, the structural distinction between the latent and the manifest. The manifest signifies the appearance of things, the symptoms; the latent points to the hidden causes, the set of possibly repressed thoughts that give rise to those symptoms. Freud introduces this distinction in *The Interpretation of Dreams* (1900). This psychology engages the work of interpretation to get at the past, the hidden, the latent.

The French philosopher, Paul Ricœur, brilliantly described this process as 'the hermeneutics of suspicion'.[7] An interest in hidden causes which lie beneath surface phenomena and cannot be immediately perceived is clearly evident in Freud. But Ricœur points out that this is a common mode of operation in the nineteenth-century sciences and social thought. Some other obvious candidates for this 'hermeneutics of suspicion' are key scientific, philosophical and political thinkers of the period: Friedrich Nietzsche, Karl Marx and Charles Darwin. 'Hermeneutics' is a term borrowed from the field of religious study in the early nineteenth century, where it refers to interpretation of biblical texts. Transposed to a more general context, the term retains a sense of reading symbols and their meanings. Freud drew an explicit connection between psychoanalytic interpretation, deciphering illegible physical and mental symptoms and the translation of Egyptian hieroglyphics, by describing the task of psychoanalysis as concerned with mental hieroglyphics.

> In fact the interpretation of dreams is completely analogous to the decipherment of an ancient pictographic script such as Egyptian hieroglyphs. In both cases there are certain elements which are not intended to be interpreted (or read, as the case may be) but are only designed to serve as 'determinatives', that is to establish the meaning of some other element. The ambiguity of various elements of dreams finds a parallel in these ancient systems of writing; and so too does the omission of

various relations, which have in both cases to be supplied from
the context. (SE XIII: 177)

The principles for reading Ancient Egyptian hieroglyphic script
had been unknown by Europeans until the Napoleonic expe-
dition to Egypt of 1798. The French officer, Pierre-Francois
Bouchard, discovered the Rosetta Stone in 1799, but when
the British defeated the French in 1801, they seized the stone
and transported it back to Britain to be housed in the British
Museum. The science of deciphering the stone's Egyptian hieroglyphics, as well as its Egyptian demotic and ancient Greek,
required finding modes of intermediary translation between
known and unknown languages, hieroglyphs and cursive script.
Various scholars were involved: the English physicist Thomas
Young made a first breakthrough, but it was Jean-François
Champollion who found the crucial key to deciphering in 1822.
Comparative philology was one of the great sciences of the
nineteenth century – a historical and structural science. It
became one of Freud's own models for psychoanalysis.

Psychoanalysis as theory, method and therapy

In an entry on psychoanalysis in the *Encyclopædia Britannica*,
Freud wrote that psychoanalysis consists of three basic components: theory, method and therapy.[8] It is a theory of sexuality,
of mind and of personality; it is a method of free association,
allowing the exploration of the inner, the psychic life; through
the interpretations of the analyst, it is also a therapy, a treatment, a talking cure. Each of these components engages a
different ideal. In seeking a theory of mental processes, psychoanalysis is grounded in the ideal of science, of episteme. It
aims to produce a theory of the world. It also strives towards
an ideal of discovery, aiming at the discovery of things in the
world. It is this ideal that leads Freud to describe himself as a

discoverer rather than a doctor. On 1 February 1900, in a letter
to his fellow doctor and friend Wilhelm Fliess, Freud wrote:
'I am actually not at all a man of science, not an observer, not
an experimenter, not a thinker. I am by temperament but a
conquistador, an adventurer, if you want it translated, with all
the curiosity, daring, and tenacity characteristic of a man of
this sort.'[9]

Freud, certainly back in these early days of psychoanalysis,
felt a greater pressure to explore the undiscovered land of the
unconscious through his patients and through himself, than
to 'cure'. But that aim of a cure was nevertheless the third
ideal of psychoanalysis. It was an effort to change the world.
Psychoanalysis, as he and Josef Breuer state towards the end of
Studies on Hysteria (1895), may promise only to turn 'hysterical misery into common unhappiness' or relief from misery.
Therapy is thus, nonetheless, imbued with power, utility, an
ability to change the world by changing people. Freud is less
invested in this last ideal and this is telling; perhaps knowledge
– at least for Freud – is more important than power.

Over the course of his lifetime, Freud produced many new
theoretical interventions, ideas and terms. Let's just quickly
look at a few. Early on, Freud described a new topographical
theory of the mind. In 1900, in Chapter 7 of *The Interpretation
of Dreams*, he proposes a tripartite structure using the image of
an iceberg: the unconscious, the deeply buried part of the mind,
lies in the invisible darkness below water, the preconscious is
below water, but visible and accessible, and the conscious or
the tip of the iceberg is where our everyday awareness resides.
In 1923, Freud replaces this with the model of the Id, the Ego
and the Superego.

Freud's theories also work at the level of dynamics and economics. Central to his framework is the idea that the mind is
at war with itself, in conflict. For Freud, there can be nothing
peaceful about our inner lives. We are at war with ourselves
until we die – that is the basic Freudian thesis about the mind.

The main agency for producing unresolvable conflict is the force of repression. Freud seeks to measure the dynamics of this conflict, crafting a theory of the energies and forces at play in the mind, modelled on the new energetics of the late nineteenth century. This conflict model has been culturally influential: the idea of 'feeling conflicted' – so familiar now and so often heard in sitcoms that it seems banal – derives from Freud. For a nineteenth-century person, it would have been a near impossible utterance. Freud's basic thesis is that symptoms are compromises between competing forces in the mind. Central here is the historical, developmental account of the mind beginning with infancy and moving into adulthood. In contrast to the topographical model, this is a kind of stratigraphical approach, uncovering the mind's psychological layers, digging down, like an archaeologist, into the past, and focusing in on childhood, the earliest, most ancient experience. There is also a further different kind of historicity at play here: it involves a narrative of the inner life, the story we tell ourselves about ourselves.

The basic units of Freudian inner life are what he often calls 'scenes' – a charged psychical locality that accrues a density of meanings over time. He speaks of scenes of infantile sexuality, of primal scenes – referring to the child's sight of sexual relations between parents, actually observed or fantasized – and already in *The Interpretation of Dreams*, of the 'scene of action' in dreams, one that differs markedly from waking life (SE V: 536). All these scenes are thick with meaning, the most crucial notion in Freudian psychology. This goes hand in hand with Freud's claim to be a rigorous determinist. For Freud, nothing happens in the mind that doesn't have a cause. Every mental event in both waking and dreaming life has both cause and meaning. Thoughts, for him, do not arise at random: the seemingly meaningless carries significance. It is this fundamental idea that leads to the doctrine of hidden meanings. Freud sets out (and sets us out) to uncover the significance of seemingly random or meaningless events in our everyday lives. The key claim is that it is always possible to uncover meaning through interpretation.

Freud's second major theme is sexuality – the overwhelming force of sexuality in human life. Perhaps this is a direction inevitable in the aftermath of Darwin, a scientist Freud greatly respected, for his methods of observation as well. Darwin's theory essentially states that the natural world is governed by *sex*, the will to reproduce. The struggle for existence requires the production of offspring. This is the fundamental rule of natural selection. Freud considers this thesis in relation to the driving force of individual life and produces an extremely controversial and hotly debated account of infantile sexuality. For Freud, there is no such thing as a period of childhood innocence, or a pre-sexual age. In fact, he contends, the sexual lives of children are much more complex than those of adults. Scientific naturalism is mobilized here to apply to sexuality: the libido, the sexual drive and all that humans do to fulfil (or escape) it are wrenched, through careful observation, out of the domain of morality and shame and turned into something to be studied scientifically. Freud does this by developing libido theory.

Let's turn now to look at the 'fundamental rule' of psychoanalysis. This lies in the method of free association: the injunction to the patient to relay everything that comes into their mind during the session. Here's how Freud puts it in his *Outline of Psychoanalysis*, which he began writing in 1938. The urgency of living and perhaps dying in Nazi-occupied Vienna pressed him to put the tenets of psychoanalysis down, as he says, 'dogmatically – in the most concise form and in the most unequivocal terms':

> We pledge him [the patient] to obey the fundamental rule of analysis, which is henceforward to govern his behaviour towards us. He is to tell us not only what he can say intention-

ally and willingly, what will give him relief like a confession, but everything else as well that his self-observation yields him, everything that comes into his head, even if it is disagreeable for him to say it, even if it seems to him unimportant or actually nonsensical. If he can succeed after this injunction in putting his self-criticism out of action, he will present us with a mass of material – thoughts, ideas, recollections – which are already subject to the influence of the unconscious, which are often its direct derivatives, and which thus put us in a position to conjecture his repressed unconscious material and to extend, by the information we give him, his ego's knowledge of his unconscious. (SE XXIII: 174)

Freud believed that free association offered up an entirely new field of phenomena for empirical science. It may at first glance sound a little strange as a concept, but free association has links to earlier 'associationist' psychology, primarily that of Locke and Hume. In the German, Freud uses the interesting expression '*freier Einfall*' for free association: this roughly translates as 'free intrusion'– it describes the process by which thoughts, whose origins are unknown, come into the mind. On the couch, everything the patient thinks needs to be said, even – especially – what is disagreeable, unimportant or frankly nonsensical. The analyst encourages the patient to obey this essential rule under all circumstances. Aside from this, analysis, that technology of two, in which the patient speaks freely, and the analyst listens acutely, has no other constraints.

It was through free association that Freud discovered transference – perhaps psychoanalysis's most important phenomenon. According to Freud, transference consists of two principal elements. The first is the idea that the infantile prototypes of one's relationships are revived in the analytic session; the transference is in fact a repetition, a new edition, of infantile scenes and relationships, particularly with parents, grandparents, carers, siblings – indeed, any family members.

The second is that these prototypes are experienced as an immediate, real relationship to the analyst. All the emotions that Freud believed rule our lives – love, hate, contempt, envy, and so on – thus become active within the analytic setting. In theoretical terms, transference amounts to the actualization of unconscious wishes within the treatment. It describes how the unconscious plays out in the psychoanalytic session. The central fact of psychoanalytic treatment, transference may also, of course, initially appear as its principal obstacle, since how the patient unconsciously experiences the analyst – as mother, father, brother, say – can prevent them from obeying the fundamental rule of free association: they censor what leaps into the mind, as they once did in family settings.

Freud first mentions transference in *Studies on Hysteria* before he has even coined the term 'psychoanalysis':

In one of my patients the origin of a particular hysterical symptom lay in a wish, which she had had many years earlier and had at once relegated to the unconscious, that the man she was talking to at the time might boldly take the initiative and give her a kiss. On one occasion, at the end of a session, a similar wish came up in her about me. She was horrified at it, spent a sleepless night, and at the next session, though she did not refuse to be treated, was quite useless for work. After I had discovered the obstacle and removed it, the work proceeded further; and lo and behold! the wish that had so much frightened the patient made its appearance as the next of her pathogenic recollections and the one which was demanded by the immediate logical context. What had happened therefore was this. The content of the wish had appeared first of all in the patient's consciousness without any memories of the surrounding circumstances which would have assigned it to a past time. The wish which was present was then, owing to the compulsion to associate which was dominant in her consciousness, linked to my person, with which the patient was legitimately concerned; and as the result

of this *mésalliance* – which I describe as a 'false connection' –
the same affect was provoked which had forced the patient long
before to repudiate this forbidden wish. Since I have discovered
this, I have been able, whenever I have been similarly involved
personally, to presume that a transference and a false connection have once more taken place. Strangely enough, the patient
is deceived afresh every time this is repeated. (SE II: 302–303)

At this point, transference is already an observed, empirical
fact for Freud. But from the 1920s onwards, it becomes the
fundamental empirical datum and theoretical framework for
psychoanalysis, the mechanism through which past and present experience can be interpreted. Through his understanding
of the psychoanalytic transference as a compulsion to repeat,
Freud also develops his famous concept of the death drive.

Transference establishes the fundamental framework of
psychoanalysis: everything that happens between the patient
and the analyst is essentially a repetition of past events. As the
analysis progresses, it may uncover the significance of something as seemingly minor as the way a patient enters the room.
After five years, the patient might suddenly say: 'I remember
the first session, I walked in and thought, I wish there was a
cat I could kick, but there isn't one. I'll just have to make do.'
This seemingly slight, unimportant fantasy or thought may be
a crucial repetition, forming one of many keys to the patient's
unconscious. The repetition framework means that the analyst
is always in the dark about what is actually going on in the session: they don't yet know what is being repeated. The process
of analysis lies in discovering the hidden forces from the past
that are being repeated in the reality of the patient's current
relationships with the analyst and with other people.

The power of psychoanalysis is played out in the transference. At first glance, psychoanalysis might seem to be a mere
exercise in remembering or describing, reporting on the past.
The analyst asks the patient to describe the inside of their mind

as if they were a spectator travelling on a train, looking out of the window. In this sense, it seems to be a descriptive process – it seems to exemplify what the philosopher J. L. Austin called the constative function of language.[10] But what happens in analysis is the opposite of description. The scene of remembering becomes the scene of action. This shift from descriptive to performative speech can be theorized according to the philosophy of language of the mid-twentieth century. Between these two people, neither of whom touches each other or moves from their respective seats or really *does* anything, speech becomes incredibly active, powerful and affective. It becomes active speech.

It's important to stress that transference as a concept or 'experimental effect' destroys the distinction between the subjective and the objective. One could argue that the analyst's entire purpose is to be an objective observer, to retain a sense of un-involvement with the patient. The analyst refrains from talking about their personal life, even from directly discussing personal feelings or responses to the patient. But the patient behaves in the exact opposite way, becoming intensely involved with the analyst. This propels the analyst who cannot help but be involved, even if at a second remove. Yet their task is to observe the relationship between themselves and the patient as it builds, to monitor their own countertransference – in order to discover the unconscious forces that drive the patient.

The patient–analyst relationship was famously explored in the classic *The Fifty-Minute Hour* (1955) by Robert Lindner, a remarkable and radical American post-war lay-analyst who evoked the many sides of the analytic process with great panache.[11] In the book's final case study, 'The jet-propelled couch', Lindner describes a patient who worked at Los Alamos, the US government laboratory for nuclear weapon research attached to the Manhattan Project. As a nuclear scientist, this patient was extremely well informed. But the patient also had a theory about the existence of life on planets in other galaxies threatening to invade earth. He felt he had a responsibility to stop this process of secret invasion. Through his fantasized communications with the other galaxies, he created maps of their planets, drew up a huge system of how their civilization worked, as well as the kinds of technologies and communication systems they used. The analyst began by humouring his patient, treating all this as associative material. They worked on maps together; every session turned into a map-reading session. Finally, they got to a point where the analyst found himself enthusiastically saying: 'But the whole point about this is that it relates to that map over there. We must put these together!' The patient stood back and said: 'You really believe this stuff, don't you?' The analyst realized that he had become psychotic on behalf of his patient. But this was a gift to the analyst. 'From the moment he became aware I was sharing – or at least appeared to share – his delusion . . . it had somehow lost its potency, and the ratifications it gave him lacked their former charge of excitement.'[12] *Here* was the transference that they could work with – although it wasn't clear who was going to be the 'real' psychotic.

The importance of this dynamic was recognized by Freud. In *The Schreber Case* (1911), he had analysed the asylum writings of a well-known judge who had been institutionalized in mid-life. Judge Daniel Paul Schreber described his experiences in *Memoirs of my Nervous Illness* (1903). These included the belief that he was persecuted by God and that he had been penetrated by divine rays which communicated to him, requested things from him, and were transforming him into a woman. Describing Judge Schreber's complex and tangled psychotic system, Freud wryly suggested that only time would tell whether psychoanalysis was more of a psychotic system than Schreber's. This question about psychoanalysis is pertinent to all other systems of knowledge: on what basis do we know the elaborated system isn't a psychotic one? In the end, psychoanalysis, like all such systems, has to confront that question.

But to return from my tangent: transference is central to understanding psychoanalysis as a methodology. This central-ity has a number of implications. For instance, although, as the Budapest International Congress held towards the end of the war makes clear, Freud was in favour of rolling out psycho-therapy to the poor through 'free clinics', he was nonetheless worried that the methodology of transference would be ham-pered if payment of some sort were forgone entirely.

Below is a striking example of transference in action from the first week of Freud's treatment of the Wolf Man, a patient who went on to become famous in the annals of psychoanalysis and beyond. The description occurs in a letter from Freud to his younger Hungarian colleague, Sándor Ferenczi.

> A rich young Russian, whom I took on because of compulsive tendencies, admitted the following transferences to me after the first session: Jewish swindler, he would like to use me from behind and shit on my head. At the age of six he experienced his first symptom cursing against God: pig, dog, etc. When he saw three piles of faeces on the street he became uncomfortable because of the Holy Trinity and anxiously sought a fourth in order to destroy the association.[13]

These are now recognizably compulsive thoughts. Freud was the first person to coin the term 'obsessional neurosis', what later became obsessive compulsive disorder (OCD).[14] Here Freud uses the term 'transferences' in the plural, implying two: the first is that the patient was convinced Freud was a Jewish swindler, out to get his money like all Jews, an all-too-common anti-Semitic slur and fantasy. The second, transference, sepa-rate from the first, is more interesting: 'He would like to use me from behind and shit on my head.' The 'he' is confusingly ambiguous: it is unclear whether it refers to Freud or to the patient. To my mind, it is most likely the patient, particularly given the fact Freud later emphasizes: the first time the patient saw the woman he fell in love with, who eventually became his wife, she was cleaning the floor with her buttocks facing towards him, ready for penetration from behind. This particular sexual relationship was the precondition for the Wolf Man falling in love with a woman. Transposed to his relationship with the analyst, the fantasy takes on a homosexual nature, drawing out the close association between penetration from behind and the anal sphere. In these images of swindled money, of 'faeces' and 'shit on the head', the dimensions of anality are brought together. Freud obviously thought this material was a good start, an interesting enough case to report to a colleague after just one week with a patient.

The science of psychoanalysis?

So, having brought you this rather 'unsavoury' material, the question remains – is psychoanalysis a science? You shake your heads. How could it possibly be a science with people blathering on about faeces? What could be less scientific? Perhaps we should ask the archaeologists, who do rather a lot with faeces these days and who, in an earlier guise as delvers into the strata of 'civilizations', marked Freud's own imagination and his metaphors for the work of psychoanalysis.

But I ask again: is psychoanalysis a science? I repeat this because it is an important question, particularly for those of us in the philosophy of science, which itself emerged as a discipline in the early twentieth century and in the same intellectual milieu as psychoanalysis. The philosophy of science is importantly concerned with the problem of demarcation – the demarcation of science from non-science – a question not many people had been interested in until that moment. In particular, it became a preoccupation of philosophical groups like the Vienna Circle, who put forward the theory of logical positivism.[15] This problem of demarcation is deeply interlinked with the larger cultural questions of Freud's time such as the science versus religion debate, and the whole matter of 'naturalism'.[16] All this comes into play in the philosopher Karl Popper's famous argument that psychoanalysis cannot be scientific because it explains *too much*; it is not vulnerable to refutation and is therefore too powerful a theory. (Popper makes the same case about Marxism.)[17]

Psychoanalysis, in many respects, leads the concept of science into a kind of crisis. To return to the question of objectivity and objective knowledge: it's clear that psychoanalysis destabilizes the prerequisite for science to be impersonal and objective. Its individual analyses, its 'laboratory work', are not performed by a third party. It thus poses the question of whether *subjective* scientific knowledge is possible, or whether science *has* to be objective to count as science? Do you have to be a 'third party', 'outside', to be a scientist? Is it, by definition, impossible to have scientific knowledge of yourself? Can knowledge produced in the first person be reliable? Or is that too odd a way to think about science? To be devil's advocate for a moment, one might argue that science is just one among many perhaps more immediately important forms of knowledge. For instance, the question of where my car keys are – not a very important query, you might say – does not entail scientific knowledge by any stretch of the imagination, though I guess one might be able to deduce it through third-party evidence. But I know where my car keys are: they're in my pocket. So, the question then spills over into a different problem: can there be a human science without subjective knowledge? Put another way, there will always be a rivalry between scientific knowledge and other forms of knowledge about human beings, between scientific and folk psychology. Folk psychology is the 'real thing', it's what you and I have and what helps us get by. It's a very sophisticated epistemic system for helping us not to be annihilated in social life. Scientific psychology attempts to avoid folk psychology and its categories.

It competes with it. Freud's view of this problem was very straightforward:

> Psychology, too, is a natural science. What else can it be? But its case is different. Not everyone is bold enough to make judgements about physical matters; but everyone – the philosopher and the man in the street alike – has his opinion on psychological questions and behaves as if he were at least an amateur psychologist. (SE XXIII: 282)

We know ourselves and other people in a psychological sense; we regard ourselves as experts in psychology. The same cannot be said for other scientific matters – say, the engineering of bridges or rockets. Most of the time, we are happy to leave those to experts.

Let me finally, in these introductory remarks, note some of the attractions of psychoanalysis that gave it cultural heft in Freud's own time. After the deaths and ravages of the First World War, its consolatory function was particularly important: it provided ways of making meaningful the painful violence of that era. Meanwhile, the emphasis on sexuality, the very ability to speak of it, certainly appealed to the young and to writers, artists, intellectuals. But even in those circles there was resistance to that appeal, not least amongst artists and writers. Vladimir Nabokov, who learned about psychoanalysis as a student in Cambridge living in Trinity Lane in 1920, wrote: 'Let the credulous and the vulgar continue to believe that all mental woes can be cured by a daily application of old Greek myths to their private parts.' Tragedy and sex make a potent combination.[18] For Nabokov, who called Freud the 'Viennese witchdoctor', psychoanalysis was all nonsense.

Perhaps Nabokov's ire is partly linked to an appeal of psychoanalysis that is close to his own craft: narrativity. On the evidence of the case histories and *The Interpretation of Dreams*, Freud was a great turn-of-the-century writer. On top of that, what he describes – neurotic misery, the way in which a child becomes and remains within the adult – becomes a fascinating life-story, perhaps even a tragic one. For the American social scientist, Philip Rieff, whose book about Freud marked an American milestone, the attraction of psychoanalysis was found in 'the excitement of a wholly interesting life', the very idea that everyone has an interesting and complex inner narrative.[19] That psychoanalysis held out the hope of a cure from neurotic misery was also undoubtedly important. Finally, a more diffuse criterion, which one shouldn't underestimate, is the draw of post-Darwinian psychology. We still live in a post-Darwinian age, one in which Darwin's theories continue to be extremely exciting for scientists and for the wider culture. What could be more Darwinian than Freud? Both thinkers emphasize sex above all other human impulses. The motor for Darwin's theory of the production of variation is sex and reproduction, which becomes the major machine for natural selection. Once Darwin is in place as a prominent scientific and philosophical thinker, the emergence of psychoanalysis seems almost inevitable. But it needed Freud to chart the territory, to explore the hidden unconscious regions, of this new science – the science he described, in that BBC recording, as both 'a part of psychology and a new method of treatment of the neuroses'.

Lecture 2

The Historical Foundations of Psychoanalysis

Portrait photograph of Sigmund Freud, 1891.
© Freud Museum London

The clinical scientist

According to his biographer and close colleague Ernest Jones, Freud stopped being young in 1900, at the age of 44.[1] But who was this man who so shaped the climate of the last century and our own? I want to begin this lecture with some biographical comments about Freud himself, before returning to his early work and the invention of psychoanalysis.

During the years 1886, when he abandoned the neurological laboratory, until 1900, Freud built up his private medical practice. Strapped for regular income, he also took up a post as director of the neurological section of the First Public Institute for Sick Children. In this hospital, where free treatment was provided to sick children from poor families, Freud developed an expertise in childhood cerebral palsy. As late as 1897, so after the appearance of *Studies on Hysteria* and when he already first moots the existence of the Oedipus complex and childhood sexuality, he is publishing neurological papers that remained standard texts for many years.

During these late years of the nineteenth century, Freud not only wrote various psychoanalytic papers and *Studies on Hysteria* (1895) with his eminent friend and colleague Josef Breuer, but also completed an important book based on prior research – *On Aphasia* (1891). As the century came to its end, the epoch-defining *The Interpretation of Dreams* (1900), a self-examination propelled by his father's death, appeared. Despite all this and the birth in rapid succession of the six children who were so important to him and whose relations with each other and their parents inevitably played into his thinking, Freud called these early years ones of 'splendid isolation'. The psychoanalytic movement had not yet begun and from that point of view, he was indeed isolated.

In the first years of the new century, he continued to be prolific. Alongside *The Psychopathology of Everyday Life* (written in 1901 and published in 1904), important new cases,

Dora amongst them, a book on *Jokes and Their Relation to the Unconscious* and the ground-breaking *Three Essays on Sexuality* (all 1905), a following was beginning to take shape around him. Between 1906 and 1914, Freud and his growing cohort of early psychoanalysts embarked on the establishment of an international movement. Now came what might be described as a 'middle-period theory' in part evident in the *Five Lectures on Psycho-Analysis* (1910) he delivered at Clark University in the United States. The correspondence with Carl Gustav Jung, already an acclaimed researcher at Zurich University and a member of staff at the internationally famous Burgholzli Hospital, began in 1906. This Zurich connection marked a major scientific alliance and of course Jung, too, went with him to America, where he was, incidentally, paid higher fees than Freud, perhaps because he was a better negotiator. Freud continued to be prolific through this time, producing a series of papers on fantasy, on creative writing and more, as well as case histories – of Little Hans, the Rat Man and Schreber.

During the war years (1914–18) Freud became involved in a number of disputes with former colleagues, prompting a period of reflection. During this time, he wrote a series of highly theoretical papers on meta-psychology, as well as *The History of the Psycho-Analytic Movement (1914–18)*. By the end of the war, Freud's status as a celebrated international thinker had been solidified. Between 1919 and 1926, he set about reforming his own theories as well as building new ones, including the love and death instincts and the tripartite structure of the mind: the Ego, the Super Ego and the Id. In 1923, he was diagnosed with cancer. Although he continued to be industrious following his diagnosis, he wrote fewer technical or clinical papers, turning his attention to more expansive cultural criticism, such as *The Future of an Illusion* (1927) and *Civilization and Its Discontents* (1930).

From brains to dreams

Now let's deal with all this in a bit more detail. Freud was born on 6 May 1856 into a lower-middle class Jewish family in the small town of Freiberg (now Pribor) in Moravia, then a Czech part of the Austro-Hungarian empire. He was the first child of his wool-merchant father and his third wife Amelia Nathanson. He was preceded in the paternal line by two half-brothers already in their twenties by the time of Freud's birth. Soon after, they emigrated to England where Sigmund twice visited them as a young man. Freud's family wanted him to be a lawyer, but his enthusiasm for nature and biology made him decide to turn to neurology and medicine. He also had an early penchant for the classics and for literature – he knew his Shakespeare in English, and his *Don Quixote* in Spanish. He could read (and speak) English, French, and Spanish, alongside Latin, Greek, and Hebrew. He also loved philosophy. At the University of Vienna, he numbered among his professors the influential German philosopher, Franz Brentano, the founder of phenomenology in its post-Kantian mode. Another of Brentano's famous pupils was Edmund Husserl, often treated as the founder of the distinction between continental and analytic philosophy.

Freud's principal university discipline was biology, in which he carried out experimental research and field work. In the late 1870s, he travelled to the Mediterranean for the first time in order to study at the Trieste marine biological research lab. His subject was the testes of the eel. Ironically perhaps, he was already asking evolutionary questions about sex, here in relation to the eels' hermaphroditic function. He was taught by the eminent physiologist Ernst Brücke, who had him looking down a microscope and examining nervous tissue. Brücke imparted his own love of natural science and the dynamic working of cells to his young pupil who, if not constrained by funds, would have liked to continue working in labs and follow in Brücke's footsteps. Freud's training at the Vienna General Hospital, what we might now call a residency, had him living on the hospital grounds and rotating through various wards, including internal medicine, surgery, dermatology, ophthalmology, neuropathology and psychiatry, which involved laboratory research in cerebral anatomy. Freud was taught psychiatry and neuroanatomy by the director of the psychiatric clinic associated with the University of Vienna, Theodor Meynert, a leader in the field, though Freud and he later had differences.

Throughout the first part of the1880s, Freud sought to make a name for himself by pursuing neurological laboratory work. He hoped this might get him a job as a research scientist and professor at a university. During these years, he also specialized in research involving new methods of staining slices of brain. In 1883, he started to study the uses of cocaine, an episode in his life that has since become infamous, in part because in good nineteenth-century tradition he also experimented on himself. Not realizing that cocaine was itself addictive, Freud's idea was to use cocaine to wean people off morphine. He introduced this method while treating his friend, the physician Ernst von Fleischl-Marxow, who had become addicted to morphine. Using cocaine as a substitute for addictive drugs and an analgesic or painkiller, Freud missed its anaesthetic properties, for which it is still used today in some eye surgeries. His friend Karl Koller picked up on these, performing the first operation using cocaine as an anaesthetic in ophthalmology. It made him world-famous. Freud was away visiting his fiancée at the time: for years after, while life was financially hard, he bitterly resented his own failure to make a breakthrough by seeing cocaine's properties as a surgical wonder drug. Instead of success, he faced the opprobrium of having misused cocaine in clinical medicine.

In October 1885, Freud travelled to Paris on a scholarship to study with Jean-Martin Charcot, then the most important neurologist in the world. His first work at the vast Salpêtrière

Hospital involved not the famous hysterics, but research in Charcot's laboratory. Here he investigated children's brains. Indeed, Freud's early publications were a series of histological and physiological research papers, and the first major publication under his own name – rather than as a translator of John Stuart Mill, or Jean-Martin Charcot, or the French medical hypnotist Hippolyte Bernheim (the latter two forming the two sides of the famous hypnotism debate) – was a monograph *On Aphasia* (1891), the speech defects that occur usually as a result of a neurological insult such as stroke. Through the 1890s, Freud became an internationally respected clinical neurologist through his hefty monograph on the hemiplegias – brain injuries causing one-sided paralysis in children.

If Freud's enthusiasm for scientific research was his first chosen path, it wasn't the one he would ultimately pursue. In 1881 he fell in love with Martha Bernays, a charming, highly intelligent young woman of excellent family: her paternal grandfather had been the Chief Rabbi of Hamburg. But Martha, by the time Freud met her, was fatherless and had no dowry. Freud too was impecunious: with his first earnings, he had his parents and sisters to support. Love, the need to get married, his own inability to make a good enough living, precipitated his decision to become a practising clinician rather than a research medic. He worried that if he waited for a position as a research scientist, he would be 40 before he would have the resources to marry – certainly far too long a span for the passionate 25-year-old Freud then was. In 'Civilized sexual morality and modern nervous illness' (1908), one of his most socially engaged essays prior to *Civilization and Its Discontents*, Freud points out the various ways in which the cultural and moral settlements of his time – the prohibition on sex before marriage, long engagements, the insistence on purity and monogamy, the refusal of contraception – are all in conflict with instinct and thus lead to nervous illness. He had reason to know.

Influential in Freud's decision to become a practising doctor was his friendship with the reputed Jewish physician Josef Breuer, a man fourteen years his senior. A kind and generous man, Breuer also loaned Freud funds. When Freud returned to Vienna from Paris in February 1886, he opened a private practice in general medicine, specializing in nervous diseases – patients with neuroses, hysteria and neurasthenia, then the new American diagnosis spreading across Europe. He married later that year and had six children over the following decade.

We know most about Freud's private life during the period 1894–1901 from an extensive correspondence (only Freud's side of which survived) with Wilhelm Fliess, a Berlin ear, nose and throat specialist. In 1895, Freud published *Studies on Hysteria* (SE II), a text co-authored with Breuer. This text made his name as a clinical specialist in nervous diseases. From 1893 onwards, theories of the aetiology or causation of neuroses had become the centre of his theoretical work. He continued to put forward bold hypotheses, some of them in print. In 1896, however, he was derailed by the death of his father, Jakob Freud. This loss led to a period of confusion and depression, a sense of private failure and crisis. During a period of self-analysis, he rejected most of the theories he had developed between 1893 and 1897. He turned his attention to the one thing that seemed still to be standing: his theory of dreams eventually published in 1900 as *The Interpretation of Dreams* (SE IV & V).

With his dream book, which opened a new century, Freud became first and foremost a clinical scientist: his laboratory, his site of observation, was the consulting room. But like many of his contemporaries, Freud remained preoccupied with causes – even though he wasn't any longer a laboratory scientist, the natural habitat for this kind of science at the time.

The theoretical framework Freud used to approach nervous diseases was basic brain physiology and neurone theory as it emerged in the 1880s and 1890s. Freud comes out of the school of Hermann von Helmholtz: from the 1840s, this consisted of a reductionistic approach to (nerve) physiology which fundamentally argued that there are only physical and chemical causes at work in the organism. Freud sought to develop an account of these processes in the brain in terms of what he called 'matter in motion', drawing on the basic reflex model of the nervous system and brain. He also explored a line of inquiry in research psychiatry, following the work of Meynert. Freud began to describe brain anatomy and physiology in minute detail, leading him to aphasia theory, then at the cutting edge of scientific psychiatry (what we would now call brain science) in the period 1860–1900. The tradition of phrenology, the study of the shape and size of the cranium as a supposed indicator of character and ability, had grown to prominence between 1790 and 1850. Though it was no longer popular in Vienna by the 1870s, it provides an important background to Freud's inquiries. In 1861, the French physician Paul Broca succeeded in correlating a particular lesion in the brain with a particular symptom in aphasia, the impaired ability to understand or produce speech, arguing therefore that this part of the brain controls speech function. This led to the development of 'localization theory' in the 1870s and 80s – a tradition Freud and others sometimes called 'brain mythology' – whose aim was to relate increasingly precise parts of the brain to specific functions.

Freud's monograph *On Aphasia* put forward both a clinical and a theoretical critique of localization. He makes the clinical observation that people who suffer from aphasia sometimes do get better. There is not necessarily a close relationship between extent of damage and severity of symptoms. It was clear, Freud argued, that people recover by transferring brain function from one part of the brain to another. In other words, the brain was fundamentally multifunctional. This aligned neatly with the theoretical principle that understood the neurone as the basic unit of the brain. Freud was one of the first to model the brain as a collection of neurones. Theoretical neurophysiology at this moment aimed to find the properties of neurones, which can then be compounded and linked together to give various kinds of functions, and Freud pursued this line of inquiry, a kind of deductive brain physiology, in his neurological manuscript, *Project for a Scientific Psychology* of 1895 (SE I). Crucially, anatomy is of no interest here; instead, the whole brain is treated as one system. We might draw an analogy with computer systems; it doesn't matter to us exactly *where* software, or where a particular database, is loaded onto our hard drive so long as it is able to perform its function. Freud had a kind of 'computer hardware' view of the brain; it exists for software to be loaded onto it and to facilitate various functions.

In the work he undertook in the 1880s as a private clinician, Freud tried out many of the most fashionable methods of the scientific avant-garde on his patients. Often enough, however, he gave up on these methods, finding that they didn't work. Freud thus fell out of love with brain anatomy. The theoretical, deductive, neuroscientific approach didn't get him anywhere – apart from some interesting speculations. In his writing, one often finds that he is making complex criticisms of practices that he has tried and tested, and found ineffective, or he was simply not very good at. This last was especially the case for his experiments in hypnotism.

Hypnotism and hysteria

Freud's early theory of nervous diseases grew out of Jean-Martin Charcot's work on hysteria. Charcot famously established neurology as a profession. His clinical descriptions of multiple sclerosis, Parkinson's disease and epilepsy laid the foundations for modern neurology. He also, importantly, studied traumatic accidents. The second part of the nineteenth century had seen a burgeoning interest in the effects of such accidents, in part occasioned by the growth in railway travel. The investigation of functional failure – that is, a failure or disorder in function for which there is no (as yet) discernible physiological cause in the patient – following an accident often led to insurance disputes that ended up in court. As a result, money poured into scientific research in this area. What emerged was the view that damage to the nervous system was not necessarily organic or observable. This became a model for later nineteenth-century theorizations of neuroses in general.

Charcot, the 'Napoleon of the neuroses', pioneered the understanding of the contrast between the functional and the organic. He argued that hysteria, which until this historical moment had been seen as a trivial woman's complaint and one unworthy of serious scientific investigation, could be studied using the experimental tools of modern clinical science. Charcot thought of hypnosis not as a form of treatment, but as a tool for investigation: he used hypnosis to *simulate* hysteria. If hysterical states could be induced by hypnosis, the influence of the mind on the nervous system could be shown. Pierre Janet, one of Charcot's important followers, put forward the view that hysteria arose due to a weakness in the patient's ability to synthesize an external event: this weakness resulted in a dissociative split in consciousness. Such work by Charcot and his followers laid the foundation for the development of modern theories of multiple personalities[2] as well as some thinking on trauma. Ian Hacking, in his book *Rewriting the Soul: Multiple Personalities and the Science of Memory* (1995), notes that the first spate of modern multiple personalities occurred between 1875 and 1900. Then diagnoses dwindled until at least the 1960s, after which there suddenly seemed to be an epidemic, particularly in the United States, during the 1980s and 1990s.

Opposite is a photograph from a book called *Iconographie Photographique de la Salpêtrière*, which provides photographic documentation of the work being done at the Salpêtrière, particularly in the hysteria/epilepsy wards. The image shows the onset of a hysterical attack, probably simulated under

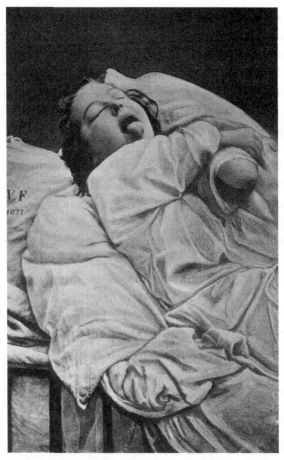

'Planche XXVIII: Début d'une Attaque', in *Iconographie Photographique de la Salpêtrière*, service de M. Charcot par Bourneville et P. Regnard 1877–80, Vol. 2. Wellcome Images

hypnosis – Charcot's technique for investigating the characteristics of hysteria. Freud took up and developed Charcot's ideas, distinguishing more clearly between organic and mental anatomy. He introduced the causal structure of trauma as the fundamental event leading to particular mental conditions. He also began to use hypnosis as therapy rather than as a research tool. In this latter usage, he followed the French neurologist

Hippolyte Bernheim, whose books – *De la Suggestion et de ses application à la thérapeutique [Suggestion and Its Therapeutic Applications]* (1886) and *Hypnotisme, suggestion, psychothérapie [Hypnotism, Suggestion, Psychotherapy]* (1891) – Freud also translated. Bernheim had criticized Charcot, arguing that simulating hysteria through hypnosis was a redundant exercise. Instead, Bernheim proposed that the fundamental force in hypnosis is a basic disposition of the nervous system or the mind to 'suggestion' or suggestibility. Both hysteria and hypnosis arise from this suggestibility; there is no autonomous psychological or pathological state. The concept of suggestion, which is a general psychological category of enormous power, was used by Bernheim, who emphasized that hypnosis should be employed as a therapeutic rather than a research tool. Mobilizing his concept of 'suggestibility', he, himself, produced miracle cures.

Now Freud didn't just translate Bernheim. He drew on his account of post-hypnotic suggestion in particular, and hypnosis in general, to develop a theory of authority on the basis of this suggestibility. In 'On mental treatment' (1890), Freud wrote that 'hypnosis endows the physician with an authority such as was probably never possessed by the priest or the miracle man, since it concentrates the subject's whole interest on the figure of the physician'. Owing to post-hypnotic suggestion, he continued, 'hypnosis enables the physician to use the great power he wields during hypnosis in order to bring about changes in the patient in his waking condition' (SE VII: 298). Freud also started to elaborate on Bernheim's categories, as we see in this passage:

> Outside hypnosis and in real life, credulity such as the subject has in relation to his hypnotist is shown only by a child towards his beloved parents, and that an attitude of similar subjection on the part of one person towards another has only one parallel, though a complete one – namely in certain love-relationships where there is extreme devotion. A combination of exclusive

attachment and credulous obedience is in general among the characteristics of love. (SE VII: 296)

One can see Freud's later work lurking beneath the surface of this interesting early observation. He notes that the pathological states that doctors encounter have a prototype in normal life. He relates the idea of suggestibility to being in love, a state during which one completely loses one's sense of proportion about oneself and the loved person and the mind no longer holds sway. Freud's emphasis on extremity here points in two directions: the pathological can be the model for the normal, or the normal can be the model for the pathological. Love can be the model for suggestibility or suggestibility the model for love (and indeed, as he later elaborates in *Group Psychology*, suggestibility in relation to subjection to an ego ideal, such as an autocratic leader). It is also evident here that Freud is already interested in developing a theory of doctor–patient relationships distinct from knowledge of specific cases. He stresses the dependence of the patient on the hypnotist, emphasizing the absolute subservience of the subject. This is part of a broader interest in investigating why patients love certain kinds of doctors and not others; why certain kinds of doctors are particularly good at faith cures, at authority and suggestibility, at what we might call placebo effects.

Freud gauged his own abilities as a hypnotist as very poor: he was not always able to hypnotize patients; and if he was, the suggestions made under hypnosis didn't hold up on waking or for any significant period. His patients were recalcitrant. But this was important, and we'll return to this point, for Freud went on to think of the analyst as needing to engage patients' resistance in order to produce an actual effect on their lives.

The case of Anna O

But let's move on now from these general remarks to the case
that marks the invention of psychoanalysis: The Case of Anna
O, a patient of Josef Breuer's, whom Freud always names as the
founder of the whole edifice of psychoanalysis.

Anna O was a young woman treated by Breuer in 1881–2.
Around 21 years old, she presented a series of incapacitating
symptoms typical of a florid case of hysteria of the period.
These included an inability to speak German, her native
tongue (although she could still speak Italian and English),
paralysis of the right arm and leg, disturbance of vision, severe
nervous cough, an inability to drink water despite being thirsty
('hydrophobia' as it was called at the time) and a series of 'twi-
light states' of confusion, delirium and hallucinations. Breuer
treated her frequently, sometimes twice a day with a morning
and evening session. He established that her symptoms were
bound up with her father's final illness through which she had
nursed him. These symptoms had begun in the aftermath of
his death.

In Anna O's treatment, and led by her, Breuer employed
the technique of storytelling. Focusing in on each of her individual symptoms, Breuer had Anna O trace its emergence
back chronologically from the present to the past until she
uncovered the first occasion on which she had experienced it.
For instance, her inability to drink water was eventually traced
back to this scene:

> One day during hypnosis she grumbled about her English
> lady-companion whom she did not care for, and went on to
> describe, with every sign of disgust, how she had once gone
> into that lady's room and how her little dog – horrid crea-
> ture! – had drunk out of a glass there. The patient had said
> nothing, as she had wanted to be polite. After giving further
> energetic expression to the anger she had held back, she asked

for something to drink, drank a large quantity of water without any difficulty and woke from her hypnosis with the glass at her lips; and thereupon the disturbance vanished, never to return. (SE II: 34–35)

Anna O's reaction of disgust to the dog drinking from her governess's glass caused her to be unable to drink water. This is a classic hysterical symptom; her identification with the governess produced the thought that it could have been her, Anna, drinking this unclean, this disgusting, soiled water. From this it followed that any water might then be dirty unless she had seen where it had come from. Her inability to drink was thus both compulsive and involved a hysterical identification, as Freud would later describe it, with the governess.

Breuer's method of returning her to the first source of a symptom and elaborating the scene was also instrumental in the treatment of Anna O's cough:

She began coughing for the first time when once, as she was sitting at her father's bedside, she heard the sound of dance music coming from a neighbouring house, felt a sudden wish to be there, and was overcome with self-reproaches. Thereafter, throughout the whole length of her illness, she reacted to any markedly rhythmic music with a *tussis nervosa*. (SE II: 43–44)

Anna's snake hallucinations and part paralysis were traced back in the same way:

In July 1880, while he was in the country, her father fell seriously ill of a sub-pleural abscess. Anna shared the duties of nursing him with her mother. She once woke up during the night in great anxiety about the patient, who was in a high fever; and she was under the strain of expecting the arrival of a surgeon from Vienna who was to operate. Her mother had gone away for a short time and Anna was sitting at the bedside with her right arm over the back of her chair. She fell into a waking dream and saw a black snake coming towards the sick man from the wall to bite him. (It is most likely that there were in fact snakes in the field behind the house and that these had previously given the girl a fright; they would thus have provided the material for her hallucination.) She tried to keep the snake off, but it was as though she was paralysed. Her right arm, over the back of the chair, had gone to sleep and had become anaesthetic and paretic; and when she looked at it the fingers turned into little snakes with death's heads (the nails). (It seems probable that she had tried to use her paralysed right arm to drive off the snake and that its anaesthesia and paralysis had consequently become associated with the hallucination of the snake.) When the snake vanished, in her terror she tried to pray. But language failed her: she could find no tongue in which to speak, till at last she thought of some children's verses in English and then found herself able to think and pray in that language. The whistle of the train that was bringing the doctor whom she expected broke the spell. (SE II: 38–39)

The model, established by Breuer and later developed by Freud, was predicated on the idea that traumatic experience leads to an abnormal state of mind which then produces autonomous symptoms. But Breuer's treatment method was arguably more important than the model. Anna O gave it an English name: the 'talking cure', or, alternatively, 'chimney sweeping'. The idea was that one might rid oneself of one's accumulated experiences by talking them out. Breuer and Freud specified that the talking cure was also a causal therapy, in that it treated the original cause – the traumatic experience – and what had been done to it by the mind in the intervening period.

Freud persuaded Breuer to publish these findings and his treatment of Anna O in a co-authored paper entitled

'Preliminary communication' (1893). Here, the two doctors put forward the thesis that 'hysterics suffer mainly from reminiscences'. The paper was immediately taken up by many physicians throughout Europe and the Americas. This is Freud's 1909 account of the founding case:

> While the patient was in her states of 'absence' (altered personality accompanied by confusion), she was in the habit of muttering a few words to herself which seemed as though they arose from some train of thought that was occupying her mind. The doctor, after getting a report of these words, used to put her into a kind of hypnosis and then repeat them to her so as to induce her to use them as a starting point. The patient complied with the plan, and in this way reproduced in his presence the mental creations which had been occupying her mind during the 'absences' and which had betrayed their existence by the fragmentary words which she had uttered. They were profoundly melancholy phantasies – 'day dreams' we should call them – sometimes characterized by poetic beauty, and their starting-point was as a rule the position of a girl at her father's sick-bed. When she had related a number of these phantasies, she was as if set free, and she was brought back to normal mental life. (SE XI: 12)

In the long case history of Anna O, which Breuer wrote for *Studies on Hysteria* (1895), he goes into considerable detail about how the patient's individual symptoms became the focus of the cure and the method.

There's a very complicated background to this story which has been much contested by historians. Anna O – or, to give her real name, Bertha Pappenheim – was in fact a close friend of Freud's fiancée, Martha Bernays. They shared a guardian when Martha's father died. Martha's uncle, Jacob Bernays, was a Jewish philosopher who had developed a medical version of Aristotle's theory of catharsis, one of the most important

Photograph of Bertha Pappenheim/Anna O.
By permission of The Marsh Agency Ltd., on behalf of
Sigmund Freud Copyrights

theories of literary criticism. Aristotle's theory contends that we turn to literature because it evokes certain emotions and, when these are experienced, say in watching a tragedy, the effect is a cleansing or cathartic one. Jacob Bernays attempted to develop a medical theory of this cleansing effect, naming it the 'Aristotelian theory of medical catharsis'. Freud and Breuer essentially transposed this theory into their account of hysteria. But Anna/Bertha also had a series of other symptoms, including many related to her morphine addiction, which did not clear up during her treatment with Breuer and sadly resulted in later bouts of hospitalization.

The accounts of the end of her treatment by Breuer are many and contested. According to Freud, the ending was very abrupt and, despite the alleviation of symptoms, basically unsuccessful. Nonetheless, Bertha became a redoubtable woman, eventually working as a respected social worker in Germany where she led a major campaign against the sexual slavery of Jewish women transported from Europe's east. She became so acclaimed in her chosen field that she was later memorialized in a DDR stamp with her photograph on it – to

Bertha Pappenheim (1859–1936), 1954 stamp

honour her status as a great benefactor of the German nation. But it was as the first person to undergo the 'talking cure' that Bertha and her case as Anna O became the starting point for psychoanalysis.

Towards psychoanalysis

As he began to assemble the building blocks of psychoana-lytic theory, Freud brought together three principal lines of thought on diagnostic, classificatory and therapeutic practice. Charcot's search for a specific description for hysteria was central to this. So was Bernheim's view that hypnosis was not a pathological state of the nervous system, but rather a general psychological means of transformation that could prove useful as a therapeutic rather than as a research tool. He also drew on Breuer's therapeutic model, in which therapy and causal inquiry exactly coincide. Freud further shaped psychoanalysis on the bacteriological model popular in the late nineteenth century; this trend in the biomedicine of the day emphasized the rigorous search for specific causes, of a set of necessary and sufficient conditions for illness. Freud's innovation was to find these sets of conditions in the complex relationship of sexual-ity and memory, the sites that would become psychoanalysis's primary domain.

Following Charcot, Janet and Breuer had developed a theory of the specific pathological states which led to the formation of neurotic symptoms in hysteria and other nervous diseases, naming them 'hypnoid' or 'dissociated' states. In a dissociative state caused by a traumatic event, all sorts of pathological mechanisms go into action – above all, splitting. This is the basic premise of the PTSD model put forward to cope with the interest group of Vietnam veterans during the late 1970s and which continues to be prominent today. This reintroduced the notion of dissociation: in a defensive action in response to external trauma, the mind could split off from actually occurring events. Conventional psychiatry continues to work with this idea of dissociation. Freud, however, rejected this mechanism: if dissociation is explained by the occurrence of a traumatic event, why do some people have a dissociative reaction and not others? And where was the psychological motive?

Freud thought that Janet and Breuer leaned too heavily on an idea of pathological heredity or disposition to explain dissociative symptoms. Nor did dissociation sufficiently account for a means of distinguishing one patient or experience from another. Here, we see Freud shifting into what we might term 'legitimate' or straightforward psychology, away from a mishmash of pseudo-neurology. He rejected the external trauma-dissociation model because it led too easily to the view that all nervous patients were deficient in a hereditary sense.

The great French observer [Charcot], whose pupil I became in 1885–6, was not himself inclined to adopt a psychological outlook. It was his pupil, Pierre Janet, who first attempted a deeper approach to the peculiar psychical processes present in hysteria, and we followed his example when we took the splitting of the mind and dissociation of the personality as the centre of our position. You will find in Janet a theory of hysteria which takes into account the prevailing views in France on the part played by heredity and degeneracy. According to him, hysteria is a form of degenerate modification of the nervous system, which shows itself in an innate weakness in the power of psychical synthesis. Hysterical patients, he believes, are inherently incapable of holding together the multiplicity of mental processes into a unity, and hence arises the tendency to mental dissociation. (SE XI: 21)

It has been suggested that Freud didn't want to go down the path of heredity, in part because of his Jewish identity.

Jews were a vulnerable group, ever accused of having suspect heredity. But Freud also sought a more complicated etiological equation in which both heredity and life experiences contribute causally to one another. He continued:

> You will now see in what it is that the difference lies between our view and Janet's. We do not derive the psychical splitting from an innate incapacity for synthesis on the part of the mental apparatus; we explain it dynamically, from the conflict of opposing mental forces and recognize it as the outcome of an active struggling on the part of the two psychical groupings against each other. But our view gives rise to a large number of fresh problems. Situations of mental conflict are, of course, exceedingly common; efforts by the ego to ward off painful memories are quite regularly to be observed without their producing the result of a mental split. The reflection cannot be escaped that further determinants must be present if the conflict is to lead to dissociation. (SE XI: 25–26)

Freud replaced the earliest framework provided by Breuer and Janet with a theory of repression and the unconscious. He hypothesized that the individual unconscious may be overtaken by pathological elements in part as a result of the abnormal operation of repression. This posed a further problem, since Freud was convinced that 'normal' repression takes place as well: repression is a normal process for working over (or getting rid of) the unfortunate circumstances we might otherwise be overwhelmed by. Crucially, however, pathological repression does not allow for such experiences to be worked through, instead preserving them in what we might imagine as the refrigerator part of the mind.

> We were led to the assumption that hysterical symptoms are the permanent results of psychical traumas, the sum of affect attaching to which has, for particular reasons, been prevented

from being worked over consciously and has therefore found an abnormal path into somatic innervation. The terms 'strangulated affect', 'conversion' and 'abreaction' cover the distinctive features of this hypothesis. (SE VII: 272)

Freud thought that the degree of repression, the force of the original response to a traumatic experience, might be measured by the patient's *resistance* to approaching that repressed element. Resistance was Freud's way of indexing and measuring repression. Let's take Freud's account here:

I found confirmation of the fact that the forgotten memories were not lost. They were in the patient's possession and were ready to emerge in association to what was still known by him; but there was some force that prevented them from becoming conscious and compelled them to remain unconscious. The existence of this force could be assumed with certainty, since one became aware of an effort corresponding to it if, in opposition to it, one tried to introduce the unconscious memories into the patient's consciousness. The force which was maintaining the pathological condition became apparent in the form of *resistance* on the part of the patient. It was on this idea of resistance, then, that I based my view of the course of psychical events in hysteria. In order to effect a recovery, it had proved necessary to remove these resistances. Starting out from the mechanism of cure, it now became possible to construct quite definite ideas of the origin of the illness. The same forces which, in the form of resistance, were now offering opposition to the forgotten material's being made conscious, must formerly have brought about the forgetting and must have pushed the pathogenic experiences in question out of consciousness. I gave the name of '*repression*' to this hypothetical process, and I considered that it was proved by the undeniable existence of resistance. (SE VII: 23–24)

Contrast this with Freud's earliest guise as a private practitioner, following Charcot and Bernheim, when he had tried out the methods of the hypnotist, the psychical healer. He had made clear then that the therapist had incredible power over the patient's mind and could perform miraculous cures (the French, in their therapeutic enthusiasm, even believed at the time that hypno-therapy could cure cancer or tuberculosis). On the one hand, then, Freud felt the therapeutic relationship was like no other human relationship; it was more powerful, even than a love or parental relationship. But on the other hand, Freud noticed the patient's resistance to the therapist's desire to investigate their mind. The more powerful the original forces involved in repression, the more powerful the resistance. These are two different models for making sense of the therapeutic relations. Freud switched from one to the other as he turned away from the hypnotic method:

> I gave up the suggestive technique, and with it hypnosis, so early in my practice because I despaired of making suggestion powerful and enduring enough to effect permanent cures. . . . Besides all this I have another reproach to make against this method, namely, that it conceals from us all insight into the play of mental forces; it does not permit us, for example, to recognize the *resistance* with which the patient clings to his disease and thus even fights against his own recovery; yet it is this phenomenon of resistance which alone makes it possible to understand his behaviour in daily life. (SE VII: 261)

Through his failures to induce hypnotic states, his patients' resistance to follow his commands while under hypnosis, Freud became interested in the obstacles to this power relationship. He noticed that even if hypnotic treatment was successful for a few weeks, the patient invariably returned with another symptom. It seemed that patients weren't as pliable as the hypnotists believed – indeed, they could at times be recalcitrant. The power relationship failed. So Freud became preoccupied with *why* patients opposed treatment. One could argue that this question became the fundamental theoretical starting point for psychoanalysis.

This interest in recalcitrance was also what made psychoanalysis distinctive. In an introductory lecture on psychoanalysis in 1917, Freud stated that the difference between psychoanalysis and hypnosis or other suggestive therapies is that these second deal only in the positive transference – when the doctor is admired, loved or in some sense deeply respected by the patient and the patient wants to get well because of the doctor. Psychoanalysis takes it as given that the positive transference should be up for discussion, not something the physician rides on the back of. But for psychoanalysis, the most important thing is negative transference. In other words, the psychoanalytic process gets interesting when the patient hates or despises the analyst or thinks she is a crook and a charlatan. That's the moment when you treat the patient most seriously and begin to unpack their resistance, to uncover the origins of their views.

This dialectical relationship that builds up between the doctor and the patient is by no means an easy one. It is full of love, hate, resistance, contempt, admiration. Freud developed this into a kind of 'thermodynamic' model of energy relations at work in the mind. Here he uses the concept of *Besetzung* in German, meaning 'investment' or 'occupation', which has been translated as *cathexis* for want of a comparable English term. In this model, Freud is beginning to work on two different levels. He describes the symptom as both a mnemic symbol produced with a psychological motive, but also a compromise-formation of mental forces at work in the mind.

To try to illustrate what Freud means by 'mnemic symbol', imagine that a person starts crying when they walk past Charing Cross station. They have no idea why. Let's suppose they have an unconscious historical memory of the origins

of the naming of Charing Cross, which followed the death
of Edward III's wife, Eleanor. She had died in the north, and
her body was brought back to London for burial. Every evening, when her procession made a stop on their journey, they
erected a cross in her honour and memory. Charing Cross,
initially 'chère reine cross', was the final memorial cross on
this route. If a person cries as they pass this site, it is as if they
are remembering that thirteenth-century event of sorrow, the
death of a queen, a memory lost in their unconscious. This is
the comparable state of mind, or psychological mechanism, for
an individual hysteric and their relationship to their symptom.

In the mid-1890s, Freud was developing a sense that the
mind had a number of further mechanisms at work. 'Different
neurotic disturbances arise from the different methods adopted
by the "ego" in order to escape from [its] incompatibility [with
an idea]' (SE XI: 122). The basic model here is that the mind
comprises different parts or elements that are incompatible, in
tension or conflict with each other. Repression is one way of
keeping particular thoughts, memories, ideas out of consciousness. It involves the *splitting* of affect and idea – a splitting that
leads to obsessions and phobias. An obsession might entail
a compulsion to commit an act without knowing the reason
for doing it. The affect, which often holds the meaning, has
been split off. Another mental mechanism Freud identified is
projection: a mental element is projected outward and seems
to be perceived independently in the world yet with designs on
the individual: this leads to the structure of paranoia. Further
mechanisms include displacement and substitution and,
finally, one that preoccupied Freud towards the end of his life,
the splitting of the ego, different from the splitting of a mental
element. Instead, the ego is split into two parts which don't
know – or prefer not to know – about each other.

Freud thus began to develop a clinical phenomenology for
the mental events described to him by his patients. He started
from the position that defence is a 'normal', essential mode of

operation of the psyche. Neurosis arises from a pathological defence operating under an abnormal condition. From this starting point, Freud seeks to understand what these abnormal conditions are. In one instance, he describes the genesis of a facial neuralgia – persistent severe pain in the face – of his patient Frau Cäcilie M.:

> When I began to call up the traumatic scene, the patient saw herself back in a period of great mental irritability towards her husband. She described a conversation which she had had with him and a remark of his which she had felt as a bitter insult. Suddenly she put her hand to her cheek, gave a loud cry of pain and said: 'It was like a slap in the face.' With this her pain and her attack were both at an end. There is no doubt that what had happened had been a symbolization. She had felt as though she had actually been given a slap in the face. (SE II: 178)

Freud identifies the meaning of her symptom, but there remains the problem of why she generates this particular symptom in this particular way. Why does she feel as though she has been slapped in the face, rather than, for instance, slapping *him*? Freud suggests that she experienced the symptom either because she repressed her desire to slap her husband in the face, or she identified with his desire to slap her in the face.

Already in 1895, Freud realized that his work was leading him in unexpected directions:

> It still strikes me, myself, as strange that the case histories I write should read like short stories and that, as one might say, they lack the serious stamp of science. I must console myself with the reflection that the nature of the subject is evidently responsible for this, rather than a preference of my own. The fact is that local diagnosis and electrical reactions lead nowhere in the study of hysteria, whereas a detailed description

of mental processes such as we are accustomed to find in the
works of imaginative writers enables me, with the use of a few
psychological formulas, to obtain at least some kind of insight
into the course of that affection. (SE II: 160–161)

Psychoanalysis begins to occupy a curious hybrid space
between the scientific and the imaginative. This hybrid space
is quintessentially evident in a passage in which Freud recalls
the examination of a midwife with some colleagues in the
1880s. They ask her the question: 'Why does the newborn baby
sometimes defecate during childbirth?' The midwife replies: 'It
means the child's frightened.' In the absence of a physiological
account, the other doctors present dismissed the midwife's
answer as ridiculous old wives' medicine, or folk psychology.
Freud reports: 'She was laughed at and failed in the examination. But silently I took her side and began to suspect that this
poor woman from the humbler classes had laid an unerring
finger on an important correlation' (SE XVI: 397).

For Freud, the midwife's answer had seemed sensible,
and grounded in experience. Perhaps, he implied, old wives'
midwifery was better than pseudo-scientific midwifery.
Throughout his life, Freud straddled the side of the midwife
and the side of orthodox science. It wasn't always an easy ride.

Lecture 3

Dreams and Sexuality

The seduction theory

In the mid-1890s, when *Studies on Hysteria* was published, Freud's principal hypothesis for the distinctive cause of neurosis is what has come to be called the Seduction Theory (1894–7). In technical ways, this is a continuation of the earlier external trauma theory, although for Freud, the trauma was always in the domain of the sexual. Breuer himself had once casually noted that 'the great majority of severe neuroses have their origin in the marriage bed'. Deploying Breuer's clinical method, Freud pushed back through his patients' memories in search of the sexual experience that produced current symptoms. It was a subject that he would later develop into a cultural critique of sexual mores in general. In these early days, he thought that a necessary (though not sufficient) condition for neurosis in adult life lay in sexual abuse or seduction in prepubescent childhood. However, there was an epistemic requirement that there be some condition for defence being pathological. In search of this, Freud was led into an investigation of the development of sexuality and its relationship to mental life.

In 1895, Freud had a key idea. Defence may be essentially directed outside the self. The mind expects traumatic experience to come from outside. However, a *pathological* defence is put into place if an upsurge of traumatic experience unexpectedly comes from inside, but is perceived by the subject to come from the outside. This inside/outside dialectic combines two fundamental themes in the history of psychiatry, namely the conceptualization of insanity as 'error' and the idea of instinct as a source of involuntary impulses that lie beyond reason. The former follows Locke's theory, put forward in *An Essay Concerning Human Understanding* (1690), that insanity is a pathological association or false connection between ideas and perceptions, a disordered reasoning. In other words, insanity is error, a false perception of the world due to an error in the mind or brain.

This shifts early in the nineteenth century towards the idea that what triggers insanity is not one's erroneous relationship to the outside world, but the welling up of passions or instincts from within. This is a founding idea of modern psychiatry. Instinct comes to be understood as that which overpowers the mind. This can be seen in some of the famous legal cases of the period. In a famous case and trial of the early nineteenth century, a young servant, Henriette Cornier, offered to look after her neighbour's 18-month-old child. When the parents had gone, she took a knife, cut off the baby's head and threw this out the window. Why would she do this? How are we to understand such an act? Mind doctors were called in to observe and to offer diagnostic opinions. Cornier sat passive, saying almost nothing. It was cases like this that led to the development of the theory of pathological instinct, a savage, unconquerable force from within that overcame reason.

Trying to balance the claims of outside versus inside, Freud gradually moves towards the latter, the instincts; in particular, he zeroes in on the sexual instinct. He establishes a notion of premature sexual experience as a necessary condition for the development of a neurosis. The classic template for this is the case of Emma Eckstein, who, when 8 years old, was sexually assaulted by a man in a shop. As Freud describes it in the early *Project for a Scientific Psychology* published only after his death, a shopkeeper had grabbed at Emma's genitals through her clothes. On Freud's account, at this early age Emma doesn't yet understand anything about adult sexuality and therefore doesn't altogether know what this act means. At the age of 13, however, when she is pubescent, she enters a shop and notices the shop assistants laughing at her, one of them leering and whispering to the other. The assistants' laughter 'arouses (unconsciously) the memory of the shopkeeper' and with it brings 'a sexual release'. This last is transformed into anxiety: she fears a new attack from the laughing shop assistants, runs away and refuses thereafter to go into shops having connected her own inner recognition of sexuality and her earlier viola-tion, the assault that came from outside in the world. Taken together these two events bring on her illness.

Freud pursues this hypothesis of the link between neurosis and early sexual abuse – abuse not altogether grasped and then revivified and turned into symptoms with puberty. Yet by 1897, Freud is beginning to think the statistics don't make sense: if his hypothesis is correct, it would mean that all his patients, female or male – including himself with the inference that his own father was an abuser – are implicated. With this theory, the numbers of sexually abused would be greater than the already large number of neurotics. One way forward would be to broaden the concept of abuse and seduction, to abandon sex in order to talk about abuse in a broader sense. But Freud doesn't go in that direction (though the pan-sexualist in Freud could be seen as a response to this problem). Instead, he dis-cards his seduction theory, in large part for a different reason: Freud comes to think – as he makes clear in his letter to Fliess of 21 September 1897 – that in the unconscious there can be no distinguishable difference in potency between what is remembered and what is fantasized. He has 'the certain insight that there are no indications of reality [*Realitätszeichen*] in the unconscious, so that it is impossible to distinguish between truth and emotionally charged fiction that has been cathected with affect. (This leaves open the possible explanation that sexual phantasy regularly makes use of the theme of the parents.)'[1]

The operative words here are 'in the unconscious'. In the unconscious there are no markers for reality, so truth and fiction cannot readily be told apart. It's worth stating here that Freud's concern with this problem illuminates his affinities with British empiricism. He was encountering a problem endemic in empiricist philosophy, one that also concerned neurologists of his time and brain scientists in ours: how do we differentiate between a memory that was once a perception, and a fiction. His complicated answer in his neurological model – which is beyond the scope of these lectures – was to use the language system as a method of secondary reporting. Freud was under no illusion: children were sexually abused. Sadly, this frequently happened. But narrated in free association on the couch, imaginative elaboration and factual remembering elide, certainly in the murky area of sexuality. The task of analysis was to work with that confusion. In a sense it was the rejection of the seduction theory that gave birth to psychoanalysis itself, with its exploration of sexual perversions alongside the memories or fantasies of perverse acts, and its location of childhood motives for defence in the love and hatred for parents. For it was this rejection that led Freud to the notion of 'unconscious phantasy' – and to his treatment of dreams, alongside memories, as pathological symptoms and compromise formations with the same unconscious causality. Freud's self-analysis, which took place in the years after his father's death in October 1896, together with his analysis of his own dreams in parallel with his analysis of patients, led him in these years to the theme of the erotic relation in the fantasy of the child with his parents, to the story of Oedipus and Hamlet and to the theory of the wish contained within the dream which acted as its fulfilment.

Dreams and the wish theory

From 1895 on, dreams become central to Freud's theory and the analysis of them to his practice. Dreams become, as he writes, 'the royal road to the unconscious'. Central to the dream is the wish. 'Reality – wish fulfilment. It is from these opposites that our mental life springs', Freud wrote in 1899 to his closest friend of those years, Wilhelm Fliess.

In Chapter 3 of *The Interpretation of Dreams*, Freud presents the most direct form of wish-fulfilment at work in dreams in the example of 'a young medical colleague who seems to share my liking for sleep':

> The landlady of his lodgings in the neighbourhood of the hospital had strict instructions to wake him in time every morning but found it no easy job to carry them out. One morning sleep seemed peculiarly sweet. The landlady called through the door. 'Wake up, Herr Pepi. It's time to go to the hospital!' In response to this he had a dream that he was lying in bed in a room in the hospital, and that there was a card over the bed on which was written: 'Pepi H., medical student, age 22.' While he was dreaming, he said to himself 'As I'm already in the hospital, there's no need for me to go there' – and turned over and went on sleeping. In this way he openly confessed the motive for his dream. (SE IV: 125)

In this example, there are two wishes being fulfilled: first, Pepi wants to carry on sleeping; second, in dreaming a hallucinatory reality of being in the hospital, Pepi gets to be in two places at once (which is, in a sense, what dreams are for).

Freud pursued his tracking of dreams even to his own children. In the same chapter in *The Interpretation of Dreams*, he includes a dream his daughter Anna had. It is one he had recounted to Fliess (Letter 73 of 31 October 1897) soon after its occurrence: in the book he compares the dream to one his mother, Amelie, had after an operation.

> My youngest, daughter, then nineteen months old, had had an attack of vomiting one morning and had consequently been kept without food all day. During the night after this day of starvation she was heard calling out excitedly in her sleep: 'Anna Fweud, stwawbewwies, wild stwawbewwies, omblet, pudden!' At that time she was in the habit of using her own name to express the idea of taking possession of something. The menu included pretty well everything that must have seemed to her to make up a desirable meal. The fact that strawberries appeared in it in two varieties was a demonstration against the domestic health regulations. It was based upon the circumstance, which she had no doubt observed, that her nurse had attributed her indisposition to a surfeit of strawberries. She was thus retaliating in her dream against this unwelcome verdict. (SE IV:130)

So, Freud's hypothesis in *The Interpretation of Dreams* that a dream is a fulfilment of a wish – which he seeks to prove with his own dream of Irma's injection – comes side by side with the intertwined thesis that meaning is to be found in dreams and that these are capable of being interpreted. This is the necessary precondition for describing dreams as wish-fulfilments:

> Dreams are not to be likened to the unregulated sounds that rise from a musical instrument struck by the blow of some external force instead of by a player's hand; they are not meaningless; they are not absurd; they do not imply that one portion of our store of ideas is asleep while another portion is beginning to wake. On the contrary, they are psychical phe-

nomena of complete validity – fulfilments of wishes; they can
be inserted into the chain of intelligible waking mental acts;
they are constructed by a highly complicated activity of the
mind. (SE IV: 122)

Step by step, Freud's dream book builds up this argument showing how dreams work through various 'textual' techniques of
distortion – through condensation, displacement and revision.
He develops his definition of a dream gradually, by working in
a quasi-philological fashion on the dream as a text. A dream,
he argues, is not only a fulfilled wish: it is a *distorted* fulfilment
of a wish. With the notion of distortion, Freud introduces the
distinction between the manifest and the latent, between what
is apparent and what is beneath the surface. When he adds
that a dream is a *distorted* fulfilment of a *repressed* wish and,
finally, that it is a *distorted fulfilment of a repressed unconscious infantile* wish, he is articulating the importance of the
continuity between infancy and adulthood and stressing the
latency period (between infancy and puberty) as the moment
when that complex relationship to the infantile is established.

Let's stand back for a moment and ask ourselves why Freud
landed in this terrain of dreams and wishes – like old wives' and
fairy tales, the stuff of common lore, breakfast recitation and
everyday speech. When Martin Luther King declared, 'I have
a dream', everyone understood that the dream and the wish
were interlaced. When Disney's Jiminy Cricket in *Pinocchio*
sings, 'When you wish upon a star / Your dreams come true',
we all nod inwardly, at least in hope. Ordinary language places
dreams and wishes so closely together they can hardly be
distinguished semantically. Freud also has another reason for
placing them so closely together: 'Thought is, after all, nothing
but a substitute for a hallucinatory wish, and it is self-evident
that dreams must be wish-fulfilments, since nothing but a
wish can set our mental apparatus at work' (SE V: 567). So
Freud determines in the section on wish-fulfilment within his

second-volume ruminations on the 'Psychology of the Dream-Processes'. Here Freud is beginning to emphasize internal over external motivation. Thinking is always motivated by the passions, which have their origins in the body. This is resonant, again, with Hume, for whom reason was always enslaved by the passions. Freud agrees that humans are driven by their passions. He turns his general psychological theorizing into a neurological model, according to which the basic unit of the passions is a wish.

This is the basis of the Freudian worldview: the world is constituted by desire. For Freud, wishing, *desire* or what he calls *libido* is the driving force of our lives, our cultural endeavours and our institutions. Contrast this with other political philosophies: for Freud, neither self-interest, nor power, nor economic rationality is foundational to social and political life, but the search for pleasure (particularly sexuality in its many different forms). The centrality of desire here has a particular effect when it comes to Freud's most critical readers: the many examples, arguments and counter arguments that make up his dialectical reasoning have the effect of turning his readers into Freudian dreamers and provoking a wish to prove him wrong. This desire becomes the best confirmation of his own theory of wish-fulfilment; in entering into a critical relationship with Freud, one enters into the Freudian universe of desire, underscoring his claim about what the universe is like.

A crucial feature built into Freud's dream theory as well as into other of his reflections – which might be viewed either as cause for criticism or grounds for reflection on what psychology makes possible – is the intermingling of the language of quantity or energy, and the language of meaning. The great French philosopher Paul Ricœur contends that Freud thus mixes two incommensurable discourses – two languages that don't belong together. It could be argued, though, that psychology actually requires two such incommensurable discourses: a commonplace language of description – of ourselves and

others – as well as a language of quantifiable science, of measurement. Freud adopts a language close to physics: he talks of
the conflict of forces, of libido, of economics of quantity, of
a charge model for neurones. Using late nineteenth-century
theorizations of electricity, he sometimes models the mind as
a series of capacities and condensers. He draws on the neurone
theory of the 1890s, a model derived from the axiomatic properties of neurones as electrical charges. But he marries this
with a psychological discourse of intentionality, of meaning,
of practical rationality, beliefs and desires. This goes back to
Aristotle's practical syllogism, the idea that beliefs and desires
lead to a course of action. According to the Aristotelian psychologists, such a model of human rationality can be seen in
our everyday decisions and acts. But what Freud adds in is the
notion that we are unconscious of our own practical rationality. This position leads to Ricœur's 'hermeneutics of suspicion',
a programme for deciphering the hidden meanings and inten-
tions beneath the surface of our acts and speech.

There are thus two Freudian languages for psychological
processes, two languages for brain and mind, just as there are,
you might say, two languages – at least – for computers: one
for the hardware, one for the software. You can talk in terms of
electromagnetic forces, in terms of quantum mechanics, or in
terms of machine codes. A software engineer trying to land a
rocket on Mars would turn to computer software engineering
as opposed to the physics of quantum states. Just so, Freud
becomes interested in the 'software' of the mind. But he nevertheless also has recourse to the language of the brain, in the
same manner that a computer engineer, primarily concerned
with machine codes and software, might at some point need
to talk about the state of the machine itself. In this way, Freud
finds it useful to retain both languages.

The Interpretation of Dreams mingles the language of meaning with the consideration of a brain model. Nor does the book
adhere to a single genre. It's an autobiography, a psychological

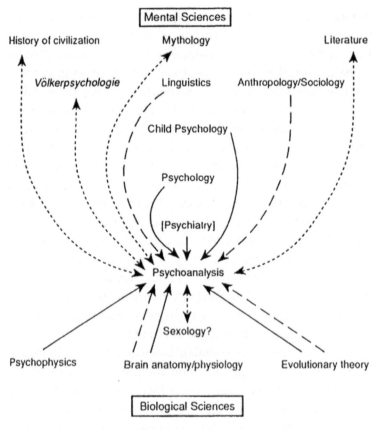

Figure 1   Diagram from Patricia Kitcher, *Freud's Dream:*
A Complete Interdisciplinary Study of the Mind (MIT: 1992), Fig 4.1
on page 111. © 1992 Massachusetts Institute of Technology, by
permission of The MIT Press

monograph, a social and cultural history of bourgeois Viennese
daily life, a 'how-to' guide to the study of dreams. It also presents a grand theory of the mind divided into the unconscious,
the pre-conscious and the conscious with two basic principles
of functioning: the primary process and the secondary process. In Patricia Kitcher's interesting book, *Freud's Dream: A*
Complete Interdisciplinary Study of the Mind (1992), which

examines Freud's strategy in relation to the contemporary cognitive sciences of the 1980s, she underlines the central feature of hybridity in Freud's work. Psychoanalysis is both a human and a biological science, refusing to choose between the two. It is also situated at the crossroads of many disciplines – as I argued in my doctoral thesis – from neurology to philology.[2] Kitcher draws up the useful diagram shown in Figure 1. Here we see that psychoanalysis sits between the biological and mental sciences. Feeding into psychoanalysis are the many disciplines that Freud was very familiar with, including evolutionary theory as well as brain anatomy and physiology – his areas of professional expertise. We see the impact of psychophysics, the experimental psychology of the second half of the nineteenth century. Kitcher sensitively puts psychiatry in square brackets, perhaps uncertain of its significance, or where it belongs – it could be argued that brain anatomy influences both psychoanalysis and psychiatry independently of one another. The relationship between psychiatry and psychoanalysis is a complex and delicate one. We also see the disciplines which influence psychology, including sexology. The dotted arrows delineate a mutual influence, flowing in both directions between psychoanalysis and literature, and the history of civilization. '*Völkerpsychologie*', or folk/social psychology, disappears as a discipline around 1920 and is absorbed into sociology and anthropology. Each of these disciplines is extremely important for psychoanalysis and must always be borne in mind when thinking about Freud's theory.

Kitcher refines this diagram to give a map of Freud's dream theory (see Figure 2). Dream theory is a crossroads for these many different disciplines. The mental sciences, literature, German neurophysiology of the late nineteenth century, as well as the influence of John Hughlings Jackson, an important English neurologist, all converge here. Freud is a magpie and draws on thinking in these various areas to produce his own theory of dreams – that 'royal road' to the unconscious. This

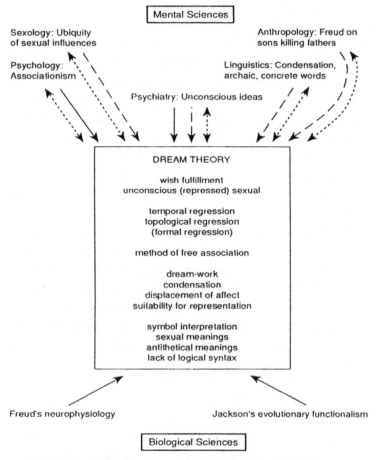

Figure 2   Diagram from Patricia Kitcher, Freud's *Dream: A Complete Interdisciplinary Study of the Mind* (MIT: 1992), Fig 5.1 on page 142. © 1992 Massachusetts Institute of Technology, by permission of The MIT Press

unconscious is a space in which – as in dreams – time does not exist, opposites or countervailing emotions (as well as passionate drives) can sit side by side, space is easily dislocated, archaic events live on, perhaps in disguised form, and logic bears little relationship to reason.

Sexuality

Side by side with the importance of the unconscious and dreams in the Freudian edifice, is sexuality – a subject that made Freud both famous and infamous. In Freud's radical conceptualization of sexuality, it's important to underscore that psychoanalysis for him is neither a biology nor an analysis of social customs: it is *psychology* and is founded on human experience.

> Sexual love is undoubtedly one of the chief things in life, and the union of mental and bodily satisfaction in the enjoyment of love is one of its culminating peaks. Apart from a few . . . fanatics, all the world knows this and conducts its life accordingly; science alone is too delicate to admit it. (SE XII: 169–170)

In this quintessentially Freudian rhetorical move, scientists are disparaged for ignoring sex as central to human life, a fact seemingly obvious to everyone else. Science here emerges as not only inhuman but obtuse. The task is to rectify or reform it, to make it recognize this basic truth about human beings – that sexual love is 'undoubtedly one of the chief things in life'. What precisely did Freud mean by this interestingly indefinite phrase?

Freud's theory of sexuality is part of a general re-evaluation of sexuality in mores, ethics and science at the end of the nineteenth and the beginning of the twentieth century. He shares this re-evaluation with several other reformers and radicals, such as Havelock Ellis and a range of bohemians and modernist artists. In his influential study, *The History of Sexuality*, Michel Foucault describes this movement as a kind of 'general incitement to speak' about sexuality; the turn of the century is the moment, for Foucault, that sexuality is fully named and required to speak its truth.

Freud's theory is distinctive for the period in its interest in infantile sexuality and in the shaping force of memories from childhood. For Freud, early childhood is the most important stage in the development of human sexuality. This emphasis stems in part from his conviction that infantile sexuality is at the root of neuroses. In a 1919 paper entitled, 'A child is being beaten,' he talks of the wild passions of childhood: he discusses there how children's first beating phantasies 'were entertained very early in life: certainly before school age, and not later than in the fifth or sixth year'. After this, Freud argues, children enter what he calls the latency period and become unreasonably reasonable. They start to feel more like adults and are capable of rational conversations. Contrast this with a child below the age of 5 or 6: an adult in conversation with a child of this age will be pulled onto their territory to the point of babbling like a baby and engaging in nonsense.

Freud explores this early childhood period of the 'wild passions', viewing it as the most important phase of life (though certainly not a time of innocence as the Victorians would have construed it). From the laboratory of his life and practice – his private practice as well as his years working with children at Vienna's Kassowitz Institute, the public and free centre for child health, only the second of its kind at the time in Europe – he develops a theory of the relationship of the infantile to the perverse in adult sexual life.

This category of the 'perverse' is central to the modern history of thinking about sexuality. Today, when much sex of all kinds is so publicly available on screens, the 'perverted' has come to refer principally to the paedophilic and abusive relationship of adults to children. Freud's discussion of the perverse is much more capacious: it reaches from sexual relations with corpses to sexual excitement from eating faeces (coprophagia). In the early twentieth century, the word 'pervert' also referred to homosexuality. These are not the 'perversions' that preoccupy the contemporary tabloid press. But our unsettlement about sexuality in relationship to children has made Freud's work on this topic a source of outrage since its inception. For one of his key contributions has been to break down the distinction between the normal and the pathological, the normal and the perverse: for Freud we are all perverts. In making this claim, he rejects the notion of childhood innocence. If children are, as he famously says 'polymorphously perverse', then the infantile is continuous with and causative of adult sexuality and character in general. He writes in 'A child is being beaten': 'It is in the years of childhood between the ages of two and four or five that the congenital libidinal factors are first awakened by actual experiences and become attached to certain complexes' (SE XVII: 184).

Freud therefore places the sexual, and infantile sexuality and the memory of childhood, at the centre of an understanding of human life. 'The sexual behaviour of a human being', he notes, 'often *lays down the pattern* for all his other modes of reacting to life' (SE IX: 198).

Sexuality thus serves as the basis and model for all interpersonal relationships. This is a sexuality that includes love and hate (or their lack), wish and fantasy, all ages and forms of expression (or their absence). Freud is also reconfiguring the relationship of sexuality and morality. Sadism and masochism take on new importance. Deriving pleasure from one's own or another's pain lands us in the domain of hatred and power. The model of love that is important to Freud here is also a distinctive one. It's crucial to remember a primary use of the word love had long been centred on Christian tradition and overlain either with piety or sentiment. When Freud talks about love, he does not mean Christian love or friendship; nor does he contrast parental and erotic love. Freud's stance is a radical one in the revaluation of values. Like Nietzsche, and in contrast to the prevailing discourse about the Victorian family, Freud repositions love and forcefully argues that parental love is not in any simple way devoted or overweening, but the purest derivative of original narcissism.

If we look at the attitude of affectionate parents towards their children, we have to recognize that it is a revival and reproduction of their own narcissism, which they have long since abandoned. The trustworthy pointer constituted by over-valuation, which we have already recognized as a narcissistic stigma in the case of object-choice, dominates, as we all know, their emotional attitude. Thus they are under a compulsion to ascribe every perfection to the child – which sober observation would find no occasion to do – and to conceal and forget all his shortcomings. (Incidentally, the denial of sexuality in children is connected with this.) Moreover, they are inclined to suspend in the child's favour the operation of all the cultural acquisitions which their own narcissism has been forced to respect, and to renew on his behalf the claims to privileges which were long ago given up by themselves. The child shall have a better time than his parents; he shall not be subject to the necessities which they have recognized as paramount in life. Illness, death, renunciation of enjoyment, restrictions on his own will, shall not touch him; the laws of nature and of society shall be abrogated in his favour; he shall once more really be the centre and core of creation – 'His Majesty the Baby', as we once fancied ourselves. The child shall fulfil those wishful dreams of the parents which they never carried out – the boy shall become a great man and a hero in his father's place, and the girl shall marry a prince as a tardy compensation for her mother. At the most touchy point in the narcissistic system, the immortality of the ego, which is so hard pressed by reality, security is achieved by taking refuge in the child. Parental love, which is so moving and at bottom so childish, is nothing but the parents' narcissism born again, which, transformed into object-love, unmistakably reveals its former nature. (SE XIV: 90–91)

In this 1913 paper devoted to the concept of narcissism, Freud argues that the best evidence for the existence of narcissism is the behaviour of parents themselves. Parents treat their children with obsessional and unbounded love; here is the proof of their own early experience of narcissism. The first months of a new baby's life shape an unconscious memory of a plenitude that may be sought throughout life, but rarely recaptured – except with the birth of the next generation. Critics of this theory might argue that Freud is here simply describing parental love. The evolutionary psychologists using the economic model would argue that what Freud calls narcissism is rather a relation of 'vestedness' in their future capital.

But for Freud, the basic model of love is the bodily satisfaction through another. This has its high point in the image of a baby with its mother: 'No one who has seen a baby sinking back satiated from the breast and falling asleep with flushed cheeks and a blissful smile can escape the reflection that this picture persists as a prototype of the expression of sexual satisfaction in later life' (SE VII: 182). This expression of satiated satisfaction on the face of both the baby and the lover are linked, for Freud, by the fact of recurrence: 'There are thus good reasons why a child sucking at his mother's breast has become the prototype of every relation of love. The finding of an object is in fact a refinding of it' (SE VII: 222). Here is the fundamental theme of Freudian psychology. Memory, bodily memory, is crucial to the making of the individual. It is clear to Freud that we constitute ourselves through memory and that we are constituted most profoundly – more than we can ever know – through *repressed* memory. An analogy from the process of language learning provides ballast for Freud's emphasis on the importance of this period, of which we have no memory. Our brains are coded for language, but if we don't learn a language between the age of 2 and 4, we never will. In this period, our language is laid down, and in some deep sense, makes us who we are. My prototype for the experience of the world is English, grounded in the fact that my mother tongue is English though I would have had equal capacity for another. I have no memory of learning English, but palpably it has made me who I am.

The next important thing to note about Freud's theory of sexuality is that sexuality is decoupled from biological reproduction. Freud implies that biology is misleading. The preconception that sex must be about reproduction obscures most of what is interesting about human sexuality. Instead, Freud is concerned with the individual's experience of sexuality: reproduction is secondary. In this sense, psychoanalysis stands in stark contrast to eugenics, a contemporary discipline that is completely preoccupied with reproduction but not with sexuality. After a century of developments in the sexual sciences, and through the contributions of feminist analyses of gender, the idea that sex and sexual pleasure and sexual orientation are distinct from reproduction and sexual orientation is now widely accepted. Writing at the beginning of this turn, Freud's interest in pleasure over reproduction has undoubtedly been influential.

When thinking about the experience of sexuality, Freud also chooses not to give any determining significance to the distinction between the sexes. He thus detaches sex from gender. His take is clearly still of its time, but embedded in it is a remarkably modern move. At the opening of Lecture XX of the *Lectures in Psychoanalysis*, 'The Sexual Life of Human Beings', he says: 'Seriously, it is not easy to decide what is covered by the concept "Sexual". Perhaps the only suitable definition would be "everything that is related to the distinction between the two sexes". *But you will regard that as colourless and too comprehensive*' (SE 16: p 303; my italics). The division between 'the two sexes' is only one dimension of the sexual. In other contexts, however, Freud disregards gender as any kind of psychically determining factor. In much of his discussion of sexuality, he also sets aside intercourse itself, instead focusing in on acts such as masturbation and kissing, and on fantasy, where love, envy, infatuation, hate and that very strong emotion, disgust, come into play to shape our practices and attitudes.

Further, Freud pointedly detaches sex from questions of morality and regulation. He urges a refusal of the equation of the sexual and the immoral. He does not view sexuality as only that which is taboo, constrained or prohibited, yet he is crucially interested in the many things humans do that at least until recently went unspoken.

Here's a further stab he makes at characterizing the sexual: 'Something which combines a reference to the contrast between the sexes, to the search for pleasure, to the reproductive function and to the characteristic of something that is improper and must be kept secret – some such combination will serve for all practical purposes in everyday life' (SE16: 304). Importantly for Freud, the workings of fantasy are the key to understanding sexuality. Sexuality is largely all in the mind; there is no such thing as a sexual act without fantasy. The patient's masturbatory fantasies are seen as in some sense key to their personality. This idea is at the centre of Freudian psychoanalytic practice: tell me what you fantasize while you masturbate, and I'll tell you who you are.

Freud developed his views on sexual fantasy by drawing on three kinds of data. His clinical investigations into the repressed wishes of neurotics in his own consulting room were foundational. He was, of course, also familiar with the sexological literature of his period; works such as Richard von Krafft-Ebing's *Psychopathia Sexualis* (1886) and Havelock Ellis's *Studies in the Psychology of Sex* (1900) offered insights into 'perverse' sexual practices – often, for the first, with the smutty bits written in Latin. Third, Freud drew on childhood material. Freud, you'll remember, was extremely well informed about childhood. Even before psychoanalysis came on the scene, he was considered a world expert on children's incurable neurological disorders such as Parkinson's disease or epilepsy. As a practising neurologist in the 1890s, he dedicated one and a half days a week to treating children at the Kassowitz Institute and wrote an exhaustive textbook on these childhood conditions.

He did not, however, psychoanalyse these children; the childhood material that forms the basis for his theories of sexuality
did not stem directly from his hospital work. Observation was
an insufficient method for acquiring knowledge about childhood sexuality. He states at the opening of his *Three Essays
on the Theory of Sexuality* (1905): 'If mankind had been able to
learn from a direct observation of children, these three essays
could have remained unwritten.' For Freud, the best way to
enter the child's mind was through the interpretation of adult's
memories and dreams. However, some post-Freudians – even
Freud's critics – have seen childhood observation as support-
ing Freud's account of the sexual lives of children. In the early
1920s, for instance, the philosopher Bertrand Russell was
characteristically acerbic about Freud's account. However, in
1929, having run a boarding school in Hampshire for a number
of years and watched and lived with children, he stated in
his book on *Marriage and Morals*: 'I will confess . . . that a
considerable experience of young children during recent years
has led me to view that there is much more truth in Freud's
theories than I had formerly supposed.'[3] In other words, the
claims that at first appear implausible, even ridiculous, have an
observable basis.

The theory of sexuality

When Freud comes to develop his theory of sexuality, he
starts by taking the commonly accepted idea that there exists
a sexual instinct, which he then deconstructs and demol-
ishes from within. He begins his *Three Essays on the Theory
of Sexuality* by pointing out that 'the fact of the existence
of sexual needs in human beings and animals is expressed
in biology by the assumption of a "sexual instinct", on the
analogy of the instinct of nutrition, that is of hunger' (SE VII:
135). Since there is no everyday word for this sexual hunger,

'science makes use of the word "libido"'. Instantly, he posits a distinction between the *object* of desire from which sexual attraction proceeds and the sexual *aim*, that is, the physical act towards which the instinct tends. He then moves into the 'numerous deviations' from what is 'assumed to be normal' in both object and aim.

Amongst the object deviations, he addresses homosexuality or 'inversion' as the most common and various. Freud talks about homosexuality in order to make claims about sexuality in general: he discusses homosexuality so as to attack the notion of degeneracy, heredity and the idea of sexuality as innate. Inversion, he says, can date back to earliest memory or to puberty, may persist through life, or constitute an episode. He dismisses the common late nineteenth-century category of 'degeneracy': he does not believe it has any value or use in respect of understanding homosexuality. Inverts, as far as Freud is concerned, are not 'degenerate', since inversion is found in people who are 'distinguished by specially high intellectual development and ethical culture'. Nor does postulating a choice between inversion being 'innate' or 'acquired' 'cover all the issues involved in inversion'. Theories of anatomical bisexuality (or the remnants thereof) being linked to inversion, or the idea of a female brain in a male body (or vice versa) can certainly not be generalized to fit all cases, especially since 'we are ignorant of what characterizes a feminine brain', while male inverts can have the most complete 'mental masculinity' or at least exhibit the same traits as their fellows. Freud never uses the category of the 'innate' with anything but reluctance. No single 'aim' can be attributed to all inverts.

In a 1915 addition to his original 1905 essay, and many patients later, Freud confesses that psychoanalysis has not yet produced a complete explanation of the origins of inversion. But he does offer some pointers as to the possible 'psychical mechanism' of its development. In the earliest years of childhood, male homosexuals may have passed through a short-lived but intense fixation on a woman, usually their mother. They leave this fixation behind, but they may then identify themselves with women and, proceeding on a narcissistic basis, take as object a young man who resembles themselves 'and whom they may love as their mother loved them'. In other words, 'their compulsive longing for men has turned out to be determined by their ceaseless flight from women'. That said, Freud adds that all humans are capable of making a homosexual object choice and have in fact already 'made one in their unconscious. Indeed, libidinal attachments to persons of the same sex play no less a part as factors in normal mental life, and a greater part as a motive force for illness, than do similar attachments to the opposite sex.' Given this, the fact that American psychoanalysts of the 1950s and '60s erred with insistent blindness in their condemnation of homosexuality appears as its own kind of perversion.

Let me repeat: in Freud's understanding, sexuality can be variable both in object and aim; both are fluid and subject to change. The sexual instinct has only a contingent relationship to choice of object:

> – a fact which we have been in danger of overlooking in consequence of the uniformity of the normal picture, where the object appears to form part and parcel of the instinct. We are thus warned to loosen the bond that exists in our thoughts between instinct and object. It seems probable that the sexual instinct is in the first instance independent of its object; nor is its origin likely to be due to its object's attractions. (SE VII: 148)

For Freud, all humans are fundamentally bisexual. His use of the term 'bisexual' is differently inflected from its modern usage, where it refers to object choice. In Freud's terms, bisexuality describes one's disposition in relation to the world; all humans carry both masculine and feminine dispositions. Thus, in his claims about bisexuality, he pushes at

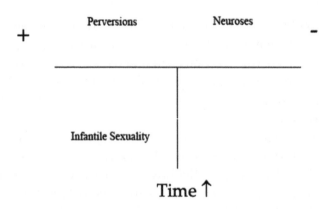

Figure 3   Freud's map of sexuality

the parameters of both gender and sexuality; masculinity and femininity mean something beyond the fluidity of choice of an object.

So, for Freud, sexuality, as well as the many aspects of desire, are acquired and fluid, traceable to initial impressions and early experience. Many famous autobiographical accounts exist in which authors pinpoint early shaping sexual experiences. Jean-Jacques Rousseau, for example, in his *Confessions* (1769/published 1789) details his experience of love at the age of 8 or 9 for a particular governess who used to beat him. He conveys the sexual excitement the beatings awoke in him. The pleasure he derived from being beaten – a masochistic relation to the loved object – became a model that took on a variety of configurations through his life. This is the kind of material Freud talks about when he talks about sexuality.

In Freud's basic model of sexuality (see Figure 3), neuroses are the negative of sexual perversions. Neurotic fantasies are shaped by that which has been repressed and defended against. The pervert instead acts out his or her fixations. In this model, sexual perverts are happy-go-lucky characters who perform socially forbidden sexual acts, despite the disapproval they entail. A dichotomy is established between the unhappy, moral, symptom-ridden and the happy, immoral sexually perverted. Both are fundamentally derived from infantile sexuality: the complex history of individual child development gives rise to it all. Various elements of the child's sexuality can split off in the course of development and turn into free-standing sexual perversions.

Freud famously establishes stages of development of infantile sexuality: the oral, anal, visual and phallic. The oral derives from the relationship to the breast, giving rise to particular inflections of character as well as to psychological defences. The infant at breast is distinguished by a desire for its incorporation and destruction, both fundamental to our continuing relationship with loved objects. Like Oscar Wilde's famous 'Yet each man kills the thing he loves' in *The Ballad of Reading Gaol*, Freud underlines how humans take the beloved into themselves: they incorporate the other, simultaneously preserving and destroying them; they satisfy a hatred for them, which is also a hatred of their own dependency; they identify with and ultimately cannibalize them, making themselves into that object. Freud determines the sexual zones of the body with an eye to the orifices, the body parts patrolled by inside/outside 'mucous membranes', as he calls them: the eyes, the nose, the lips, the ears, the anus, the genitals, the urinary apparatus. The crossings between inside and outside are what determine the hypersexual zones of the body.

For example, on kissing, which he likes to distance from romantic understandings so that its full perversity emerges, Freud writes: 'The kiss . . . between the mucous membrane of the lips of the two people concerned, is held in high sexual esteem among many nations . . . in spite of the fact that the parts of the body involved do not form part of the sexual apparatus but constitute the entrance to the digestive tract' (SE VII: 150). What is Freud doing describing a kiss like this? He is mimicking medical terminology, almost as though trying to make the reader nauseated. He wants to show the close proximity between excessive pleasure and excessive disgust in our bodily relations – both to other people, and to an extent, to ourselves. He reminds us that sex and disgust are closely linked. At the time, kissing was socially accepted if done in private. Oral sex, on the other hand, evoked disgust:

> The use of the mouth as a sexual organ is regarded as a perversion if the lips (or tongue) of one person are brought into contact with the genitals of another, but not if the mucous membranes of the lips of both of them come together. This exception is the point of contact with what is normal. (SE VII: 151)

Why should this be the case? If sex is defined by genitals, and kissing is permitted, surely oral sex should be a socially sanctioned practice. Yet early twentieth-century European culture was repulsed by acts involving both the mouth and the genitals. Freud underscores that the mapping of perverse responses and feelings of disgust seems to be arbitrary. This argument is not new: in 1580, Michel de Montaigne compared the cannibalistic rituals of the South American Indians with the European tradition of the eucharist. He argued that the Europeans are more disgusting because they eat their God every Sunday, putting Jesus Christ into their bodies, whereas the South American Indians engage in cannibalism very infrequently and under duress. Freud makes a similar denaturalizing move in his thinking about sexuality.

This oral stage in which a kind of cannibalistic identification is performed, providing the recipe for a satisfactory love relationship, is followed by the anal – the first properly social dimension since it is marked by an extension into the outside of what has been inside.

Freud's exploration of anal-sadistic sexuality is famous. Central to his theorization is the fact that faeces are the first object for the child that is detachable and distinguishable from the body. In his 1917 paper 'Transformations of instinct in anal erotism', Freud elaborates:

> For its faeces are the infant's first gift, a part of his body which he will give up only on persuasion by someone he loves, to whom indeed, he will make a spontaneous gift of it as a token of affection; for, as a rule, infants do not dirty strangers. (There are similar if less intense reactions with urine.) Defecation affords the first occasion on which the child must decide between a narcissistic and an object-loving attitude. He either parts obediently with his faeces, 'sacrifices' them to his love, or else retains them for purposes of auto-erotic satisfaction and later as a means of asserting his own will. If he makes the latter choice we are in the presence of defiance (obstinacy) which, accordingly, springs from a narcissistic clinging to anal erotism. (SE XVII: 130)

Freud points to the often rigid organization of faeces, especially in a family setting, with potty training, nappies, washing, and more. This may well be the site in which the child experiences, his first 'no', his first prohibition. Indeed, children learn about boundaries, limits, the pleasures (and pains) of rebellion, through the gifting and withholding, the regulation of their faeces. They may develop secretive scatological practices that persist in various forms. Included in the world of faeces is the world of social and spatial regulation.

Indeed, the anal-sadistic character is someone preoccupied with the rules governing faeces. (These are also linked with the death element, with loss, regulation, and more, which I won't go into here.) Obsessional neurosis, which we now call OCD, often has an anal-sadistic component, the emphasis on cleanliness and washing being closely correlated to the tidying or control of excrement, which can later also metamorphose into a hoarding of money.

What I want to underline here is that Freud's theory of infantile sexuality is firmly built around the body and the different capabilities associated with particular organs: the mouth, anus, eyes, genitals. Children's early experiences, associated with what he terms the 'component instincts', are disparate and not yet organized into one precise constellation. It is only at puberty, when the genitals start to behave differently, that they come to be clustered around the ruling principle of the genital organs.

Freud posits that each of the organs is closely linked to children's speculations about birth. He calls their hypotheses, 'The sexual researches of childhood'. The question 'Where do babies come from?' is one that often first took the form: 'Where did this particular intruding baby come from?' – this horrible little brother or sister that has just arrived to turn my universe topsy-turvy. This first major question drives what Freud posits is the child's 'epistemophilic' instinct – a curiosity, a desire for knowledge propelled by libido, crucially bound up with the body and sexuality. The drive to know the world is, for Freud, propelled by these first avid queries about babies. Children's research here is often lonely: they know that the answers they are given are nonsensical or evasive or incomprehensible. This, Freud posits, thus becomes a first and early occasion for psychical conflict. Even if the parental answer is framed in adult terms of vaginas, wombs, penises, sperm, such official language is opaque to the child evoking nothing comprehensible. Children therefore rebel by generating their own sometimes secret theories, translating the information into their own experience of the body, perhaps amplified by fairy and folk tale. They might imagine for example, that everyone, both boys and girls, has a penis, a general enough contention amongst little ones, or that babies are born out of the anus, a place that might already have yielded some interesting sensations for them; or they posit some other space inside the body, perhaps attached to the navel. Their speculations take them far afield.

If babies are born through the anus, then a man can give birth just as well as a woman. It is therefore possible for a boy to imagine that he, too, has children of his own, without there being any need to accuse him on that account of having feminine inclinations. He is merely giving evidence in this of the anal erotism which is still alive in him. (SE IX: 218–219)

If much has been made on Freud's insistence on the phallus – as the marker of sexual difference, as the sign of power and patriarchy, or the site of women's 'lack' and so-called 'penis envy' – it's worth pausing to see just how crucial his concentration on childhood sexuality has made the 'anus' in his thinking. The anal, as I've mentioned, is the site of the first prohibition to a child. (Psychically, this is the first evocation of what Freud later calls the superego.) Faeces represent everything that is repudiated, something smelly and repellent, that has come out of the child's body. Children who see or hear parental intercourse, and its accompanying sounds, often associate these last with pain, and the organ in play as the anus, which in turn can lead to sadistic theories of intercourse. The disgust that faeces and their smell generate – the very word 'pooh!' and its accompanying facial expression – can lead to secrecy and shame, to a reaction formation in relation to the child's original instinctual impulses. The attempt to control the faeces and the attached emotions set up various character traits such as parsimony, cleanliness, orderliness. Freud also asks whether the atrophy of the sense of smell – 'an inevitable result of man's assumption of an erect posture' – and 'the consequent organic repression of his pleasure in smell may not have had a considerable share in the origin of his susceptibility to nervous disease'. His thoughts on anality and excreta thus lead him onto the terrain of an evolutionary account of the peculiar vulnerability of humans to neurosis. These are further galvanized when, around 1930, he is considering the structure of our civilization and its discontents. In a footnote to the beginning of section four of *Civilization and Its Discontents*, he writes:

> A social factor is also unmistakably present in the cultural trend towards cleanliness, which has received *ex post facto* justification in hygienic considerations but which manifested itself before their discovery. The incitement to cleanliness originates in an urge to get rid of the excreta, which have become disagreeable to the sense perceptions. We know that in the nursery things are different. The excreta arouse no disgust in children. They seem valuable to them as being a part of their own body which has come away from it. Here upbringing insists with special energy on hastening the course of development which lies ahead, and which should make the excreta worthless, disgusting, abhorrent and abominable. Such a reversal of values would scarcely be possible if the substances that are expelled from the body were not doomed by their strong smells to share the fate which overtook olfactory stimuli after man adopted the erect posture. Anal erotism, therefore, succumbs in the first instance to the 'organic repression' which paved the way to civilization. The existence of the social factor which is responsible for the further transformation of anal erotism is attested by the circumstance that, in spite of all man's developmental advances, he scarcely finds the smell of his own excreta repulsive, but only that of other people's. Thus a person who is not clean – who does not hide his excreta – is offending other people; he is showing no consideration for them. And this is confirmed by our strongest and commonest terms of abuse. It would be incomprehensible, too, that man should use the name of his most faithful friend in the animal world – the dog [bitch] – as a term of abuse if that creature had not incurred his contempt through two characteristics: that it is an animal whose dominant sense is that of smell and one which has no horror of excrement, and that it is not ashamed of its sexual functions. (SE XXI: 100)

Here Freud is also providing an explanation for the basis of social organization. Disgust and repulsion provide a kind of warning about other people. For Freud, disgust tells you something about social relations. These emotional reactions offer a way of diagnosing social structure. Yet the point of society – 'civilization' – is to organize the human body and patrol it through disgust and shame – just think of the world-historical significance of the water closet, its historical import in class and status, in defining the 'ranking' of a country. What sexuality does is to override that disgust, while narcissism goes some way in redefining its limits.

If I've paused overlong on Freud's speculations about the childhood phase of anality and its sequelae, it's because we are much more familiar with his other stages. The eyes, the visual, the stage that follows the anal, are of special import. This visual stage gives rise to several typical sexual perversions, including exhibitionism, voyeurism, forms of fetishism, such as the later interest in high heels. The cultural primacy of the visual and its links to sexuality are most spectacularly obvious in the fashion, film and pornography industries. The genital phase, with its focus on the penis and clitoris, takes the child into masturbatory terrain and leads Freud into a theory he elaborates more fully in the 1920s as applicable to both sexes – the castration complex. 'Both male and female children', he writes in a 1920 footnote to 'Infantile sexuality', 'form a theory that women no less than men originally had a penis, but that they lost it by castration. The conviction which is finally reached by males that women have no penis often leads them to an enduringly low opinion of the other sex' (SE VII: 195). It is certainly open to debate whether this, along with the very fact of sexual difference and the boy's earliest dependence on the mother, results in some of the excesses of misogyny; or indeed in girls being 'overcome by envy for the penis – an envy culminating in the wish, which is so important in its consequences, to be boys themselves' – an aspect of Freud's reflections that has been hotly debated since the early days of psychoanalysis, and more vigorously with second-wave feminism.

Let's stand back a moment and take stock of how this account of sexuality fits in within Freudian theory as a whole. As we've seen, Freud's theory of sexuality unfurls within his larger model of the mind, which incorporates a conflict theory and a theory of instincts. From the 1890s through to the 1930s, Freud always posits a conflict of some sort between the self (which is not altogether a stable entity) and its formation by conventional morality and the sexual instinct. Take an example: in an early case history in *Studies on Hysteria* (1895), the young woman he calls Elizabeth von R (Ilona Weiss) can't acknowledge that she has fallen in love with her dead sister's husband and converts this conflict into a hysteria; she develops pains in her legs and difficulties in walking. Freud unearths, through cathartic treatment and concentration, a link to her father in his last illness, when he used to rest his leg in need of bandaging on hers. But her resistance to getting better sends Freud in search of repressed material, which he finds in an erotic wish, a desire, linked to her sister's husband. He is heard next door, and Freud notes the sharp pain his patient then suddenly experiences in her leg. Here the self and the sexual are at odds with each other. Though Elizabeth still morally and mentally resists Freud's sexually charged interpretations, the very expression of them (and her mother's confirmation thereof) soon results in her dancing again, marrying and having a child.

The conflict model persists in various other ways in Freud's work. In the *Three Essays on the Theory of Sexuality*, various conflicts between the ego and the sexual mark or disturb development. These conflicts between the self and the sexual had resonances with nineteenth-century biological metaphors, with their distinction between nucleus and protoplasm, between the individual and the species. In Freud, the conflict between ego and self took the form of psychology; conflict was staged between self and object, hunger and love. He later sees three different players in his vision of conflict: the mind, the body and the world. What is clear is that, for Freud, sexuality bases the mind in the body and is in turn the source for wishes. The mind is ultimately driven by the demands made on it by the body (rather than the world). The secondary process, thinking, is an inhibition and defence against the primary process, the unconscious wishes, which come from the connection that the mind has with the body.

From 1910 onwards, Freud embeds his theory of sexuality in his dynamic theory of instincts, developing these ideas through his account of instinct – which he frames as 'a concept on the frontier between the mental and the somatic'. In 1915, he writes:

> If now we apply ourselves to considering mental life from a biological point of view, an 'instinct' appears to us as a concept on the frontier between the mental and the somatic, as the psychical representative of the stimuli originating from within the organism and reaching the mind, as a measure of the demand made upon the mind for work in consequence of its connection with the body. (SE XIV: 121–122)

For the purposes of psychoanalysis, Freud designates the instincts as '*Trieb*' or drive: these drives are 'innate' mechanisms, in that sense biological, and shared with the species. They are human hard-wiring with external triggers. Eros – or sexual passion, desire for a sum of excitation, the propeller of creativity – is joined in 1920 in *Beyond the Pleasure Principle* by Thanatos, a competing death instinct born out of Freud's observation of a human compulsion to repeat, to engage in destructive and self-destructive acts, plus a tendency for all systems to return to a quiescent state.

Thus the Freudian mind is shaped and misshaped by con-flicting desires and drives that stem from the body, itself both biological and individual. In turn, their trajectory within the self is moulded, inhibited, forgotten, repressed, distorted, defended against, translated by morality and the social and cultural settlements of an epoch. Being human is never, for Freud, an altogether simple matter.

Lecture 4

Psychoanalysis as a Theory of Culture

Portrait photograph of Sigmund Freud, 1914.
© Freud Museum London

We're going to shift the topic a little this week to see how Freud translates his thinking about individuals and their inner lives – dreams and the unconscious, the constitution of the self through sexuality and repression – into the social and cultural sphere.[1]

It's sometimes said that, because of his reluctance to make political proclamations, Freud was only interested in the sphere of the psyche – some would say only the psyche of the wealthy who could afford private consultation. Psychoanalysis as a whole has often been criticized for its lack of political engagement – its focus on navel-gazing. But Freud, despite his pessimism, was not disengaged. The Fifth International Congress of Psychoanalysis, held in Budapest in 1918, two months before the armistice, marked the end of a barbarous world war, and hotly debated war neurosis, a subject that fed into Freud's *Beyond the Pleasure Principle* (1920). Freud's own address to his assembled fellows was a radical rallying call, thirty years before the creation in Britain of the National Health Service, for making state-funded and thus free mental health available to all. The poorest man, he urged, should have as much assistance with his mind as he now has with his body. Soon after, in 1919, in 'Lines of advance in psycho-analytic therapy', he notes:

> The conscience of society will awake and remind it that the poor man should have just as much right to assistance for his mind as he now has to the life-saving help offered by surgery; and that the neuroses threaten public health no less than tuberculosis, and can be left as little as the latter to the impotent care of individual members of the community ... institutions or out-patient clinics will be started, to which analytically trained physicians will be appointed, so that men who would otherwise give way to drink, women who have nearly succumbed under their burden of privations, children for whom there is no choice but between running wild or neurosis, may be made

capable, by analysis, of resistance and of efficient work. Such
treatments will be free. (SE XVII: 167)

In the wake of Freud's inspiring conference address and such
invocations, a radical younger generation of analysts idealisti-
cally set up free clinics for the poor in Berlin, Vienna, London
and elsewhere.

Practical action apart, you might say that, as a scientist of
the turn of the twentieth century, a person of authority in
that broad German category of *Wissenschaft* – a word that
encompasses science, research, scholarship and knowledge
– Freud had a duty to extend and generalize his findings, as
well as his methods of interpretation. Through this optic,
Freud can take on the aura of a 'moralist', in Phillip Rieff's
mid-twentieth century sense.[2] He is a thinker who presents us
with a philosophical vision of life, one who introduces a new
understanding of the self, a self that is hidden, in conflict with
itself, prone to the eruptions or disfigurements of desire, and in
whom self-vigilance is ever in play. This self-vigilance extends
to a suspicion of the other: their authority, their own reliability
and that of their narratives, all of which call for interpretation.
Freud's vision is also at core a therapeutic one, and his picture
of 'psychological' man certainly has an important role in forg-
ing a 'therapeutic' twentieth century in which religion is losing
its social force. There is, however, a question about whether it
provides an alternative to morality as well.

Yet, if Freud doesn't immediately step into what we now
consider to be Darwin's scientific mantel, he is also not altogether comparable to other 'modern intellectuals', like Russell
or Sartre, Arendt or Foucault. Given that Freud's ideas also
form the basis of a practice and find their way into the formation of an age of therapy, the social effects of his ideas percolate
in very different ways. When you conjure up Freud, you also
imagine an institutional apparatus of professional 'followers',
who provide treatments. Many of these last treatments may

no longer be classed as so-called classical Freudian therapies, but they nonetheless bear a distinct relationship to psychoanalytic practice as he conceived it. He is a founder of a form of expertise – that talking and listening treatment of two which Freud launched on the world – and this places him in a distinctive category to other classic thinkers in modern social and political thought.

As we have seen, Freud's early scientific training and practice were neurological. His medical-psychiatric antecedents lay in psychopathology. But his own explanatory ambitions were broader than both of these and from very early on ranged outside the clinic and consulting room. He wanted to provide a theory of both individuals and society that would have the authority of a science. He also wanted to provide an account of the foundations of society, morality and authority, and to answer the difficult question that, after the First World War, became increasingly pressing: what claim does civilization have for our respect given its fundamental injustice? Investigating pathology, considering its constitution and treatment became a way of also garnering greater knowledge of the 'normal' mind and 'normal behaviours' of culture and society as a whole.

Towards a universal theory: dreams, literature and myth

Freud's focus on dreams is a case in point and a key index of his aspirations to construct a universal explanatory social theory. Dreams are inherently universal. Everyone dreams. An explanatory theory of dreams thus promises universality, encompassing everything and everyone. Dreams have been the subject of discourse for millennia, regarded in many societies as sacred, even as messages from the gods. By the end of the nineteenth century, the special status of dreams and their explanatory potential can be caught in the great English neuropsychiatrist John Hughlings Jackson's phrase: 'Tell me everything about dreams and I'll tell you everything about insanity.'

In his search to understand the associations of the individual patient and to provide a universal explanation for human behaviour, Freud doesn't only study dreams. In the first years of the new century, he adds everyday jokes and slips of the tongue (Freudian slips, as they came to be known) to his repertoire to develop through these universal cultural and behavioural phenomena a generalized account of the unruly unconscious. Myths and stories become part of his explanatory resources too: like jokes and dreams, these are shared. By treating each association, whether in a patient or in a text, in the same way, Freud further breaks down the distinction between the 'normal' and the 'pathological'. Indeed, he starts to use his studies into pathological phenomena as a basis for understanding 'the normal'. The psychopathologist becomes, in Freud's hands, the critic of the social in general. He accumulates further evidence around common associations, by drawing on the work of late nineteenth-century cultural scholars in the fields of *Völkerpsychologie*, social psychology, anthropology, religion, folklore studies, sociology and literature. All these feed into his theories of culture.

We have already seen how Freud stretched his explanatory framework from the pathological to the normal when dealing with sexuality and the development of children in his *Three Essays* of 1905. In 1908, he extends his thoughts on anal erotism into a generalized analysis of the anal character and its configuration. He describes the anal character as parsimonious, obstinate and clean – in reaction to the 'dirt' of faeces. This disposition, which one might identify as the impeccable bourgeois character in much mid-twentieth-century literature, derived for Freud from the vicissitudes of anal sexuality.

Also in 1908, in *Civilized Sexual Morality and Modern Nervous Illness*, Freud provides an important account of the conflict between the social and the sexual arrangements of Western civilization, a problem that had long concerned him, even before his *Three Essays on the Theory of Sexuality*, and was to take its strongest form in his perennial best-seller *Civilization and Its Discontents* (1930). He argues in 1908 that culture and civilization are built on the suppression of instincts: each individual has surrendered 'some part of the sense of omnipotence or of the aggressive or vindictive inclinations in his personality' as well as a part of their sexual instincts – often under the sanction of religion. The consequence is that 'all who wish to be more noble-minded than their constitution allows fall victims to neurosis; they would have been more healthy if it could have been possible for them to be less good' (SE IX: 186–187, and passim). Abstinence, the inhibition of ordinary sexuality, the policed or self-policed narrowing of its range – by lack of education, by double standards for men and women, by monogamy (which often also enforced a long period of chaste engagement between the parties), by lack of contraception and a taboo on so-called 'perversions' – Freud argues, all this leads to the nervous illness so many doctors from America to Austria have variously diagnosed. The suppression affects the entirety of lives. In marriage, women are most afflicted and often face 'the choice between unappeased desire, unfaithfulness or a neurosis':

> The cure for nervous illness arising from marriage would be marital unfaithfulness. But the more strictly a woman has been brought up and the more sternly she has submitted to the demands of civilization, the more she is afraid of taking this way out; and in the conflict between her desires and her sense of duty, she once more seeks refuge in a neurosis. Nothing protects her virtue as securely as an illness. (SE IX: 195)

Freud here is attuned to the injustice of sexual morality, sug-
gesting that modernity demands a new social-sexual settlement
if individuals are to thrive.

As a well-educated, Austro-German bourgeois scientist,
Freud had always been interested in literature and the arts.
But in the first decade of the 1900s, he makes an intriguing
theoretical foray into literature, an extension of his varied
assertions that literature, its dramas and plots and psychology,
had long preceded the work of analysts: imaginative writers
had always known about the unconscious and its workings. In
Delusions and Dreams in Jensen's 'Gradiva' (1907), he analyses this short, highly popular novel first serialized in Vienna's
principal newspaper, the *Neue Freie Presse*, in 1902.

Gradiva is the story of Norbert, a young archaeologist working in Pompei, who, in the heat of the midday sun and amidst
the petrified dead of the buried city, suddenly sees a Roman
statue of Gradiva, a young woman, come to life. The statue
talks to him and he falls in love. It gradually becomes clear that
the statue is in fact a woman he knew in his youth, a woman
he has failed in his overwrought state to recognize, lost in the
dusty layers of his own long-buried past. In Pompei, Norbert
has hallucinated this repressed relationship and transformed
his conflicted love into a statue come to life. His old, returned
flame now raises him from the dead dusty work of archaeology:
offering renewal and affectionate care, she gives him a chance
to come back from the grave by falling in love.

Freud mobilizes this story as an allegory for psychoanalysis,
drawing parallels between the role of the young woman and
the analyst. Quite explicitly here, he makes the case for psychoanalysis as a cure through love – through the phenomenon
of transference, the fantasized love of the patient for the analyst, which is, as you'll recall, in analytic terms, a new edition
of older relations through which the patient can revisit past
knots, fixations, patterns. But the difference in Jensen's story,
Freud points out, is that the real Gradiva is able to return the

young archaeologist's love. In psychoanalysis, this mutuality is not feasible: the analyst cannot enter into substantive love relations with a patient. Nevertheless, there are clear parallels between the novel's basic ingredients and the psychoanalytic method; the recovery of childhood passions, present-day symptoms and hallucinations from the past, buried psychological relics come back to haunt life. The novel grapples with these in a way that mirrors Freud's practice of analysis. Freud, in turn, analyses the novel in strictly psychoanalytic terms, now open for all to grasp in his interpretation. *Gradiva* also has the bonus of taking place on a dig. Freud, like many in his times, was fascinated by archaeology and its excavation of the ancient world; indeed, he grew into an ardent collector of ancient objects. Archaeology served as one of his favourite parallels for the kind of digging into the buried individual past psychoanalysis undertakes.

During this period, Freud further elaborates psychoanalysis as a theory of culture by investigating the creative writer's literary pursuits and finding an originary point in them in child's play and daydream: at its most fundamental level before aesthetic questions are confronted, these are way of ordering the world according to our wishes and phantasies. Myths, Freud hypothesizes 'are the distorted vestiges of the wishful phantasies of whole nations, the *secular dreams* of youthful humanity' (SE IX:152). It is in the Oedipal *drama* with its riddles, incestuous desires and murderous impulses to kill or displace the father, that Freud finds an analogy for the life, the fears and wishes, of the developing child in an ordinary family.

Just a little later, in 1910, Freud turns to Leonardo da Vinci to make a foray into psychobiography, despite his own reservations about the project of biography as a whole. In a letter to Jung, he explains that the spur to the enterprise came from a patient who reminds him of Leonardo, though without the talent – a patient who, like Leonardo, never finishes things:

Do you remember my remarks in *The Sexual Theories of Children* ... to the effect that the children's first primitive researches in this sphere were bound to fail and that this first failure could have a paralysing effect on them? Read the passage over; at the time I did not take it as seriously as I do now. Well, the great Leonardo was such a man; at an early age he converted his sexuality into an urge for knowledge and from then on the inability to finish anything he undertook became a pattern to which he had to conform in all his ventures: he was sexually inactive or homosexual. Not so long ago I came across his image and likeness (without his genius) in a neurotic. (17 October 1909, letter 158)[3]

Freud centres his analysis on a rare memory that Leonardo records from childhood.

There is, so far as I know, only one place in his scientific notebooks where Leonardo inserts a piece of information about his childhood. In a passage about the flight of vultures, he suddenly interrupts himself to pursue a memory from very early years which had sprung to his mind: 'It seems that I was always destined to be so deeply concerned with vultures; for I recall as one of my very earliest memories that while I was in my cradle a vulture came down to me, and opened my mouth with its tail, and struck me many times with its tail against my lips.' (SE XI: 82)

Freud plumbs this in order to offer up an interpretation of Leonardo's intense relationship with his mother and engage in what is an early discussion of narcissism – as a 'cool repudiation of sexuality – a thing that would scarcely be expected of an artist and a portrayer of feminine beauty' – and the nature of sublimated homosexuality. This biographical essay on Leonardo, if not Freud's finest work, since it includes a number of mistakes, shows psychoanalysis stretching its wings out of the clinic and into the culture as a whole and Freud using a different set of resources for his inquiries.

This period between 1907 and 1913 is the time of Freud's friendship with the younger Swiss psychiatrist and psycho-analyst Carl Gustav Jung. The rivalry between the two men, however formative the exchange of ideas, grows palpable in the years leading to their rupture. Jung was particularly ambitious when it came to his desire to apply his psychology to the whole of culture: he drew a wide range of myth and folklore into the orbit of his psychology, posited a collective unconscious, found the workings of myth and legend within the individual (psy-chotic) patient, and sought to generalize back to the cultural realm. Though their styles of reasoning, patient analysis and attitudes to religion and magic, let alone to the sexual/libidinal question, were markedly different, Freud, during the years of their friendship and perhaps in competitive spirit, offers up more universalizing mythological interpretations, while also focusing in on the symbolic weight of certain images – for example, the phallus as it appears in dream and myth. Freud had long been interested in anthropology and prehistory, the 'horror of incest' and what he called in writing to his old friend Fliess back in 1897 'endopsychic myths'. But it was in this period that he really develops his account of a universal symbolism, which had been absent from *The Interpretation of Dreams* – the phallic symbols, the birth symbols, the body symbols, the associations of particular dreams (for instance, the association of insects with children). Here, Freud's grander explanatory ambitions, his search for a universal theory, becomes clear.

These ambitions are their most visible in the use he makes of Sophocles' Oedipus plays to install into the everyday lexicon of the twentieth century one of psychoanalysis's basic paradigms – the Oedipus complex. In a footnote added in *Three Essays* in 1920, Freud states: 'The Oedipus complex is the nuclear com-plex of the neuroses and constitutes the essential part of their content' (SE VII:162–163). Under the aegis of Oedipus, the child's development within the conventional family is marked for the boy by a conflict with the father over sexual desire for the mother. For the girl, a simple version installs a desire for the father and a conflict with the mother. Freud thus now claims that the Oedipus complex marks a universal crossroads for the formation of character in childhood.

In *Totem and Taboo* (1912–13), Freud takes the Oedipal conflict into the social and ethnographic field, using contemporaneous studies to generate a theory of the origins of civilization and religion. Freud wants to show what psychoanalysis can say to ethnography. He uses the findings of psychoanalysis to address some of the key questions of the social psychology and anthropology of his time (rather than in the manner of Jung using the latter to provide patterns for the former). Freud asks: Why do certain tribes organize themselves in groups according to totems, in particular totem animals? Where does this come from? How do networks of prohibitions, or taboos, take shape? Freud is particularly interested in the taboo on incest.

This period of ethnography and anthropology saw a preoccupation with the origins and uses of mythology. What do myths refer to? What is the Hebrew Bible really telling us when it conjures up the image of the flood? Is this mythology, or does the flood refer to something real? These questions had crucial implications for the development of scientific disciplines – in the last case of geology. In the early nineteenth century, geology becomes recognizably Darwinian before Darwin only when the biblical flood is put to one side and dismissed as myth. But there are also schools of thought in this period that see myths as representations of astral events – in the Comtean positivist manner, myths are understood as early attempts at scientific explanation of events in the world, for example, the rising and setting of the sun as explained by Apollo pulling it with his chariot across the heavens. Or perhaps myths, as others posit, are distorted accounts of historical events.

Contemporaneous with Freud is the Durkheimian school of sociology or anthropology. In his wonderful book, *The Elementary Forms of the Religious Life* (1912), the influential French sociologist Émile Durkheim argued that myth and religion are representations of social relations. Durkheim makes the fundamental claim: God is society. Behind the worshipping practices of society, or the explicit worshipped objects, is the social itself. His thinking lies at the origin of the functionalism that dominates so much sociology and anthropology in the twentieth century. The function of myths, religion or any similar institution is to bind society together, to make it work effectively. This is summarized in the French phrase '*conscience collective*' – translated as either collective consciousness or conscience, since the French don't distinguish between the two.

It is within this intellectual context that Freud makes the argument that myths, totems and taboos are representations of repressed familial relations. In response to those who see myths in relation to historical events, he argues that myths are also institutional relics of a universal history of mankind. This is the basic framing for *Totem and Taboo*. Even if Freud's anthropology and that of his time have been eclipsed and the language of what he calls the 'primitive' has rightly been rejected, there are still things of interest in the essay.

Taboos, in Freud's thinking, exist to establish prohibitions 'mainly directed against liberty of enjoyment and against freedom of movement and communication'. Prohibitions persist and are repeated because the awakening of a memory of the forbidden action is ever linked 'with the awakening of an impulse to put that action into effect'. The transgressor acquires the characteristic of himself being prohibited as if the 'dangerous charge' of the taboo had been transferred to him. If Freud is interested in all this, it is because, as he points out, 'some of the 'moral and conventional prohibitions by which we ourselves are governed may have some essential relationship' to earlier societies. The fear with which taboos are surrounded and which keeps them alive, their puzzling origin, their seeming lack of motive correspond to the prohibitions obsessional patients set up for themselves, complete with a 'moral conviction that any violation will lead to intolerable disaster'.

For Freud, all totems and taboos, ancient and modern, are essentially derived from the primordial totem, the father, and the underlying taboo of incest – a prohibition against marrying/engaging in sex within a narrower or wider kinship group. Freud finds a parallel in the child's way of thinking, this small creature who wants to be big and who sometimes feels omnipotent, in his polymorphous sexuality, his earliest Oedipal desire for the mother and the prohibition of that desire by the father (or their stand-ins); as well as in the replay of such desires in his patients' associations and analysis.

In a prefiguration of the concerns of *Totem and Taboo*, Freud had already written 'Obsessive actions and religious practices' in 1907, tracing how religious practices and obsessional neuroses map onto one another. Obsessional neuroses, he argues, are like a private, secretive religion, while religion is a public obsessional set of rituals. Freud thus assimilates obsessional with ritual practices – both are carried out conscientiously, must not be interrupted, and result in qualms of conscience and anxiety if neglected. The differences between them – for example, the ostensible symbolic significance of religious practice and the seemingly senseless repeated acts of what we now call OCD sufferers – disappear in the consulting room when the true meaning of obsessive actions is revealed. This last is usually found to be derived from the intimate, often sexual, experiences of the patient and the first prohibitions, an early repression of an instinctual impulse that has left unconscious guilt behind. In *Totem and Taboo*, Freud pointedly links this to the child's desire to touch his genitals. This meets with a prompt external prohibition, which may pause the act but does not abolish the instinct. Since the instinct is rarely altogether

repressed, it is experienced as temptation. The instincts continuing force, whatever ceremonials are put in place to defend
against temptation and its anticipated ill effects, sets up an
unending conflict in the individual. Obsessive symptoms and
actions are a compromise with the warring forces of the mind
and so always also bring with them, if only in the form of a
generalized excitement, some of the pleasure they are designed
to prevent. The obsessional is trapped in ambivalence: they
desire and detest the same act simultaneously. Their symptom
gradually takes on the aspect of an uncomprehended replica of
the original prohibited act.

Religion too, for Freud, is based on the prohibition and suppression of instinct, both sexual and aggressive. It sets up a
simultaneous fear of the father/deity and a hope, based on
that fear, of what the father can provide. Like the pious man,
the obsessional patient both lays hope in and fears the power
of his own thoughts. Thinking or praying might make things
come true; thought and act appear as one. In the failure to
distinguish one from the other, guilt arises. In his unconscious,
the obsessional feels he actually did murder the father or the
person he felt so angry with. The thought happened. Freud
transfers this idea of the omnipotence of thought to an analysis
of magic, with its obvious connections here to religion. In this,
his account of magic is not dissimilar to the animistic theories
of the late nineteenth century, which were particularly present
in English anthropology at this time.

Totem and Taboo culminates with a speculation about the
original generation of civilization and culture. Freud ends the
essay with a quote from Goethe's *Faust*: 'In the beginning was
the deed.' With this ending, he is affirming that the realm of
fantasy studied by psychoanalysis has bases in real history. He
has generated a parallel between the development of the child
and the development of the species. Ontogeny recapitulates
phylogeny. He has taken the basic hypothesis, which owes
something to Darwin, that humans originally lived in 'primal

hordes' or loose groupings. These were structured, Freud suggests, around a leading father or dominant male, who 'possessed' the women. At a key moment, the sons band together and murder the father. Their new group is formed at the totem meal – the eating of the father after his murder. But the sons retreat from power and a take-over of 'his' women, aghast and guilty at their actions, afraid another man will rise to win over the father's prerogative that they have usurped. Instead, they hand power over to women, and under matriarchy a religion based on the father emerges. In Freud's speculative account, the first 'rule-governed' system is matriarchy. This is the religion of primal guilt.

Having noted the various murderous attempts by sons or bands of brothers against fathers, kings or gods – figures of authority all – who are successively displaced or replaced in myth by sons and the variety of sacrificial rituals, Freud concludes that the son's (collective) sense of guilt alongside the son's rebelliousness never become extinct. Like everyone at the time who is interested in such subjects, Freud follows Frazer's lead in *The Golden Bough* (1911) in his thinking on Christianity. For Freud, Christ, who stands alone, sacrifices himself to redeem the 'band of brothers' from original sin; in so doing, a son religion grows to displace a father religion. Each communion is in that sense not only the sign of a brotherhood in Christ, but 'a fresh elimination of the father', a repetition of the guilty deed. Freud, like so many of his contemporaries, has a lot of fun pointing out that this forms the basis for Christian communion – an idea notably put forward by Montaigne and Robertson Smith. An equation is being made between monotheistic and so-called 'primitive' religions, with the suggestion that, to this day, a cannibalistic murderousness is preserved at the heart of religious practices.

Freud's rather perverse thought that guilt precedes morality is crucially important. It runs contrary to the conventional idea that transgression against an idea of right or wrong must precede guilt. Instead, Freud argues that morality is invented *after* the feeling of guilt. His claim is that the incest taboo is the *result* of the original guilt over the murder of the father. The Oedipus complex, this individual psychic formation, is thus actually the residue of a prehistoric event that lies at the foundation of civilization and persists within us in the form of primal fantasies. The important theoretical background here is the neo-Lamarckian formula (following the French zoologist Jean Baptiste de Lamarck): ontogeny recapitulates phylogeny. The Lamarckian mechanism – the inheritance of acquired characteristics – allows for traces of group experience to persist in the individual through the form of primal fantasies. History can be re-awoken; it is hardwired congenitally into the human mind.

With this venture into anthropology before the First World War, Freud stakes a claim to the grand explanatory ambitions of psychoanalysis. After the war, he develops these into a theorization of group psychology, in other words, a social theory. Freud does not think it's possible to have a group psychology that is not methodologically individualist and grounded in the psychology of individuals. In *Group Psychology and the Analysis of the Ego* (1921), he advances his theory of the ego ideal through an exploration of how groups are constituted around this ideal, embodied in a leader who stands in for the internalized and idolized father of first childhood. Groups, he says, come into being through a narcissistic identification with others. Narcissism is at its strongest, Freud suggests, when I love myself in the form of something I would like to be, an imagined ideal self that is better than I am. Those people I most admire are those I would like to be. They have some kinship with me but are different, higher. The force of Freud's argument is to point out the instinctual resources upon which groups are based. In this, Freud again joins a growing interest in the late nineteenth century in examining pathological states of mind as the source of the power of crowds. This is psychology and psychopathology coming together to generate sociology.[4]

Freud had long been interested in questions of power (and its deformations) whether within the consulting room, the developing child, the couple in love, or the family, that nucleus of society. Back in 1890, he had noted the analogy between the relationship of patient to hypnotist and the patient's love for an esteemed authority figure. It was not unlike the relation of love and authority between parent and child, or lover and beloved, and was more powerful than the relation to the priest. How power is achieved and authority maintained is, of course, a question central to political theory and social organization. The political philosopher Thomas Hobbes gives us a famous and paradigmatic view in *Leviathan*: 'All mankind [is in] a perpetual and restless desire for power . . . that [stops] only in death.' Without restraint, life would be the 'war of every man against every man' and 'solitary, poor, nasty, brutish, and short' (Part 1, *Of Man*, Ch. 13). In the Hobbesian tradition, authority is founded on the fear of war, the fear of the state of nature. Freud, in contrast, advances the idea that people give up their sovereignty out of love, not fear: effectively, you fall in (narcissistic) love with the leader.

It is tempting to read this as a reference to the rise of Hitler, though this can't really be the case since Freud was writing in 1920–1. He is, however undoubtedly interrogating the rise of organizations whose power is founded on the adoration of an individual. The discourse of group psychology originally emerged in an attempt to explain the French Revolution and the power of the mob, the crowd. But Freud is less interested in explicitly political groupings, focusing instead on the army and the church. He develops a model of how these institutions function through an external object, a leader, who corresponds to the ego ideal – the person you would like to be. The individual egos all identify with each other and overcome their narcissistic differences in relation to this external object. In doing so, they gain the power of solidarity, of being 'at one'. This can be felt at the football ground, in protest groups and politics. In such settings, one is overtaken by the immense power of the crowd – a power that comes from giving up one's differences to other people when in relation to the external object, the leader.

Freud thus mobilizes psychoanalysis as an experimental method for investigating the sources of authority. One could argue that it is tailor-made for such a project; the psychoanalytic method is predicated on the exploration of the relationship of authority between the analyst and the patient. At the very least, the analyst might ask, why does the patient keep turning up day after day for this treatment, which is so painful, so incomprehensible? What are the power relations at play here? The transference is precisely the study of this relationship of power, as well as of the patient's resistance.

In this way, psychoanalysis offers not only a method for studying authority relations in general, but the specific epistemic authority relations of science and knowledge. A range of disciplines pride themselves on being critical and reflexive – on including oneself in the account of power and knowledge relations – and psychoanalysis is one of the very first of those disciplines. It is inherently reflexive; the knowing subject, the observer, must, from a Freudian perspective, include themselves in any account and be critical of the power relations at stake in the production of knowledge. Intriguingly, this is mobilized in some of the first applications of psychoanalysis; it is used as a kind of error theory or bias testing, taking dream analysis as the model. For instance, in education, it was argued that teachers, not children, should undergo analysis, to avoid the teachers having pathological relations with children which were driven by their own unresolved Oedipus complexes. Similarly, one of the great legal theorists of the early twentieth century, Jerome Frank, argued that every judge should be psychoanalysed before being allowed to practice. Without analysis, the judge risks acting out irrational responses to criminals, imposing upon them irrationally conceived rationalized judgements. Psychoanalysis, on this view, emerges as a purification of the professionals. The authority figures – not the children or the criminalized – are rid of unconscious proclivities by means of a reflexive practice.

In 1923, Freud developed his tripartite structure of the mind, positing the existence of the ego, the id and, perhaps most importantly for his extension into the social world, the superego. The concept of the superego certainly owes something to the First World War: what was this war but righteous murderousness? Thinking about the millions who killed in the name of God, country or civilization, and the millions who died as a result, Freud becomes increasingly interested in instinctual behaviour, not only at the level of repressed sexuality. The conceptualization of the superego is an attempt to crystallize the compulsory or punitive passion of morality, as itself instinctually based. The superego becomes a critical agency: 'One part of ego sets itself over against the other, judges it critically, and, as it were, takes it as its object', as he put it in 'Mourning and melancholia' (1917). This superego bosses one around, prohibits the unruly id, censors and censures the often timorous and uncertain ego. At the same time, it functions as an ideal. The superego in me is all my values and ideals that have come into being since childhood. It becomes an oppressive weight, a heavy obligation to live up to, an unattainable ideal. In measuring oneself and others against this ideal, one turns one's murderousness both against oneself and against any who fall short. Freud calls this the sadistic relationship of the superego to the ego. In this new model, the id is the source of passions, the cauldron of semi-biological desires. The ego – which is in a sense what psychoanalysis communicates with – is confronted with conflict from both sides. The superego bullies it in the name of civilized morality, while the id presses its desires and drives. In the meantime, the reality of the world makes its own demands. Freud now portrays the ego as an embattled, weak institution as T. S. Eliot underscores in 'Burnt Norton' in *Four Quartets* (1936–42), though doubtless he also had other things in mind: 'Humankind cannot bear very much reality.' When Freud develops his theory of the death instinct in *The Ego and the ID*, he calls the superego an extreme pathological form, a 'pure culture of the death instinct' (SE XIX: 53).

Religion and civilization

Towards the end of the 1920s, Freud wrote two popular books: *Future of an Illusion* (1927) and *Civilization and its Discontents* (1930). The first is a provocative discussion of religion, infamously contending that it is either a delusion or an illusion that draws its force from the helpless child's initial dependence on a father who seems all-knowing and all-powerful. *Future of an Illusion* can be regarded as a Voltairean text, in which Freud sets himself up as a spokesman for modern science in its ongoing battle with religion. His essay has a direct parallel in Enlightenment critiques of religion in the name of science. Freud declares that he wants to save psychoanalysis and science from the priests, who he worries are making a comeback.

So Freud poses the challenge: why do we believe religion's manifest absurdities? His anti-religious answer provides a psychoanalytic account of the original dependency on the father. But throughout the book is the staged conflict between science and religion. Part of the implicit background to this can be found in the famous Scopes Monkey Trial of 1925, in which Tennessee biology teacher, John Thomas Scopes, backed by the American Civil Liberties Union and defended by the great Clarence Darrow, was taken to court by the state for teaching Darwinian evolution, in violation of the Butler Act. This was lobbied for by the head of the World Christian Fundamentals

Association and passed in March of that year. The Act made it illegal to teach Darwin and evolution in state-funded schools. Scopes was found guilty and fined, though, on appeal, a technicality had the fee rescinded. The Butler Act itself was repealed in 1967, though the anti-Darwinian stance of the Creationists is still felt in school biology today. So Freud's *Future of an Illusion* was written in response to a resurgence of what we now recognize as fundamentalist Christianity with its literalist reading of the Bible and its war on science. It's useful to note that the word 'fundamentalism' in more or less its contemporary sense came into use in 1911 at the Princeton Seminary to describe a particular Protestant sect that organized through the 1920s. With this context, *Future of an Illusion* bears a marked contemporary flavour.

The matter of religion comes up again at the very start of Freud's most renowned work of social theory, *Civilization and its Discontents*. Here, Freud takes up and pushes further many of the themes he developed in the twenties. Civilization, he argues, is built up at the expense of the individual. But given this, he asks, is there value in that for the individual? Is the individual justified in giving up their chances of satisfaction in life for civilization? Freud draws a map of the ways human beings have of surviving reality – and he treats religion in this context. He notes that, in *Future of an Illusion*, he hadn't given space to an important psychological basis for religion: the 'oceanic feeling'. This feeling is part of the well-loved baby's first experience of the world. It is a narcissistic sense of plenitude, of timelessness and wholeness in which inner world and outer feel at one, united in blissful peace. These sensations can later move into the domain of religious feeling, an apparent unity between the ego and something external and bigger, something timeless, which seems to overcome some of the toll of suffering that reality inevitably introduces into life.

In the following passage, Freud has just finished addressing the pleasures that individual pursuits such as art can give, by turning the imagination momentarily away from reality and suffering. But art cannot make us forget 'real misery':

> One can try to re-create the world, to build up in its stead another world in which its most unbearable features are eliminated and replaced by others that are in conformity with one's own wishes. But whoever, in desperate defiance, sets out upon this path to happiness will as a rule attain nothing. Reality is too strong for him. He becomes a madman, who for the most part finds no one to help him in carrying through his delusion. It is asserted, however, that each one of us behaves in some one respect like a paranoiac, corrects some aspect of the world which is unbearable to him by the construction of a wish and introduces this delusion into reality. A special importance attaches to the case in which this attempt to procure a certainty of happiness and a protection against suffering through a delusional remoulding of reality is made by a considerable number of people in common. The religions of mankind must be classed among the mass delusions of this kind. No one, needless to say, who shares a delusion ever recognizes it as such. (SE XXI: 81)

For Freud, civilization, along with its characteristics – its emphasis on order, cleanliness, beauty and intellect – is built at the expense of the individual's freedom and his search for instinctual gratification (whether in pleasure, that force of Eros which is directed outwards, or in the deadly aggressive tendencies, which also finally abut on the self). There are benefits to this curtailing of the instincts, of sublimating desire and redirecting its power into everyday affection, work, the arts, science. Society allows for the satisfaction of the pleasure principle. But the conflict between the individual and social norms, laws and repressive social control nonetheless persists, as does the burden of primal guilt and the toll of neurosis. Civilization requires inhibitions, and the repression of sexual instincts gives rise to individual neuroses, symptoms, and guilt-feelings. The superego, bearer of internalized convention, may be a good citizen counselling resignation to the status quo and its strictures and acceding to its demands. Certain satisfactions may undoubtedly lie there. But in its more sadistic manifestation, the superego may make individual life unbearable.

So how is the individual to survive the curtailments imposed by reality and civilization, with their inevitable toll of suffering? Freud drily maps possible routes towards a modicum of happiness. Apart from the delusions of religion, the pleasures of sublimation and the appreciation of beauty, there is solitude: it's quite clear for Freud that other people with their competing desires are sources of unhappiness for the ego. So, solitude and solitary work may provide some satisfaction to inevitable frustrations. Apart from that, there are chemical intoxicants – from wine to drugs – which cut the individual off from reality and can act to provide temporary and very effective relief. Finally, Freud comes to what is perhaps the most important, though not necessarily the most efficacious, 'technique' in his lexicon of the art of living: love.

I am, of course, speaking of the way of life which makes love the centre of everything, which looks for all satisfaction in loving and being loved. A psychical attitude of this sort comes naturally enough to all of us; one of the forms in which love manifests itself – sexual love – has given us our most intense experience of an overwhelming sensation of pleasure and has thus furnished us with a pattern for our search for happiness. What is more natural than that we should persist in looking for happiness along the path on which we first encountered it? The weak side of this technique of living is easy to see; otherwise, no human being would have thought of abandoning this path to happiness for any other. It is that we are never so defenceless against suffering as when we love, never so helplessly unhappy as when we have lost our loved object or its love. But this does

not dispose of the technique of living based on the value of love
as a means of happiness. (SE XXI: 82)

Love, for Freud, is what really counts in the struggle to appease
civilization's discontents. But far more than the satisfactions of
the aesthetic or the intoxicant, it is unreliable, unpredictable
and perhaps unsustainable. Freud, in this last decade of his
life, is nothing if not a wry and realistic scientist, unwilling to
provide assuaging answers to the difficulties of human life and
society, though still intent on testing the challenges it throws
his way.

Lecture 5

Psychoanalysis as an International Movement

Portrait of the attendees of the Psychoanalytic Congress at Kurhaus
Bad in Bad Homburg, 1925. Including (l. to r. row 1) Moshe Wulff,
Paul Federn, Oskar Pfister, Ernest Jones, Karl Abraham, Max
Eitingon, Sandor Ferenczi, Otto Rank. Moving up you can spot:
Anna Freud, Alix Strachey, Sandor Rado, James Strachey, Melanie
Klein, Eduard Hitschmann, Wilhelm Reich, Maurycy Bornsztajn, J.
H. W. van Ophuijsen, Franz Alexander, Karen Horney.
Photographer: T. H. Voight. © Freud Museum London, where a
fuller skeleton key is available.

To journey into an undiscovered inner world is the novel promise of psychoanalysis. Freud's is the most private of new lands to be uncovered.[1] Yet, it retains in Freud's conjuring a self-declared lineage with the great Age of Discoveries of the Renaissance; occasionally, too, it acquires the flavour of colonization and the taming of Wilderness – 'Where id was, ego shall be. It is a work of culture – not unlike the draining of the Zuider Zee' (SE XXII: 80). This New World is private, secret and interior, all the while retaining a sense of strangeness, since the Freudian past is – as L. P. Hartley noted in *The Go-Between* (1953) – a foreign country. This lost, foreign world, however, belongs not just to the occasional poet but to each and every one of us: it's a universal prehistoric psychological legacy, complete with primal fathers, phallic mothers and wonder at the mystery of sex. Inherent to all of us, this unconscious world is the democratic potential at the heart of psychoanalysis. But this democracy seems to have its gatekeepers. The counterpart to the project of the indefinite riches of the 'wholly interesting life'[2] that psychoanalysis offers the individual is the figure of the late twentieth-century analyst: sober and conservative, entirely uninteresting in the inner life he or she leads, a blank screen, so ill-defined that it has proved infinitely tempting to novelists and filmmakers as a face-savingly flexible plot device. For the patient, analysis offers a Socratic journey, infinite depth, passion and complexity; in contrast, the analyst offers up to the patient and to the ambient culture an unruffled surface, uncannily calm and infuriatingly deficient of engagement in life's passions. With these two figures, one would not expect there to be such a thing as a society of analysts, let alone an international movement.

But a movement there has been, ever since Freud convened four like-minded physicians at his home one evening in 1902. To some, this movement is akin to a secret society, a kind of Freemasons of the Unconscious, you might say; to others it is a monolithic oligarchy of conservative-minded bureaucrats.

Viewed from afar, it emerges as a principal ideological wing of American hegemony during the Cold War. From close up, it's an ordinary professional society regulating its members in a spirit that, at best, rigorously enforces high, intentionally conformist standards of scientific and professional activity, and, at worst, is a direct betrayal of its own inspiration from the ever unruly unconscious. Even amongst Freud's most enthusiastic followers, the movement has never elicited enthusiasm. Yet without it, Freud's place in the twentieth century would be more akin to Nietzsche's or to Havelock Ellis's, to Wittgenstein's or to Pierre Janet's. Without the movement there might be explorations of the unconscious, but there would be no agencies of the unconscious. There might be psychoanalysis, but there would be no psychoanalysts – and given, as we have seen, the centrality of the transference and clinical practice to Freud's vision of psychoanalysis, it's unclear what that would have looked like.

But how, as the French philosopher Jacques Derrida asked, did 'an autobiographical writing [which is what Freud's self-analysis and its record in *The Interpretation of Dreams* is] in the abyss of an unterminated self-analysis, give birth to a worldwide institution?'[3] Or to move one step away from the dreaming and writing Freud to the practising Freud: how do we leap from the basic furniture of psychoanalysis – two people, a couch, an armchair, a door that closes, a clock on the wall or in someone's pocket and an appointments book – to the psychoanalytic societies, the training institutes, an international association, the psychoanalytic presses and journals, volumes of correspondence between Freud and a multiplicity of fellows in which ideas are exchanged, the many works of art, film and literature inspired by psychoanalysis? Some might think of all these last as the secondary derivatives of psychoanalysis. Yet to imagine that psychoanalysis is in its purest form Freud's self-analysis or the meeting point of patient and analyst (inevitably what it is from the patient's point of view) is a little like buying into the fantasy that science consists in the confrontation of the researcher with nature, alone, and without any other support than the tools of the scientific trade he plies, his test tubes or seismographic readings.

To take its place in the world as a science or a practice or even a set of lasting ideas (try to imagine Kant's longevity without the institution of the university), psychoanalysis needed some kind of institutional framework. Institutions, after all, are means of preserving structures beyond the individual or individuals who occupy them. They are ways of creating symbolic bodies and minds beyond the corporeal ones of individuals. From another perspective, institutions are also the means of cultural reproduction: they are ways of achieving within culture what sexual reproduction achieves within nature. Freud, who had his first suspicion of cancer in 1917 and from 1923 on underwent thirty-three operations, was certainly aware of his vulnerability and perhaps imminent death. It's clear that he didn't want this to coincide with the demise of psychoanalysis. Whatever his own early ambivalence and worries about institutions and the formalization of psychoanalytic training, Freud got what he wanted.

The international dissemination of psychoanalysis

So, what was – what is – the international psychoanalytic movement? What kind of a movement was it? Was it more of a scientific movement or a political one, more like the Communist International or the Christian Science Movement? Psychoanalysis has bred many kinds of schools and associations – Lacanian, existentialist, and the many varieties of psychotherapy that spread across the globe. But the formal psychoanalytic movement took shape in the era of fledgling institutional internationalism after the First World War, when the League of Nations, among others, embedded a novel kind of international organization. But it also owed much to the internationalism of science in earlier eras – an internationalism propelled first by the Medieval European Universities, which used Latin as a universal medium, then the scientific societies and academies of the seventeenth and eighteenth centuries, and later the disciplinary societies and national scientific institutions of the early nineteenth century. The modern psychoanalytic movement was made possible by the proliferation of scientific internationalism before it.

The late nineteenth and early twentieth centuries marked a time of scientific professionalization and internationalization, following the invention of the modern university a century earlier. If the dominant 'carriers' for professional science at the time are still doctors, not always themselves the first to be in favour of science, this is also the period when a significant number of university, state and industry posts become available to professional scientists. From the 1830s on in Germany, the laboratory is the key training system. But it was the period of 1850–1914 that marked a new internationalism in the universities. This is the period when doctoral degrees were introduced – physics and physiology amongst the first – as well as an informal postdoctoral system, of the kind Freud took advantage of in travelling to Charcot's Salpétrière. Laboratories and universities were very gradually being opened to women and minorities, such as Jews. In Vienna, Paris, Cambridge and especially at the University of Zurich, international students, particularly women, gathered in new research centres and formed bonds that could endure and be revivified in later international meetings and congresses. In Freud's early days this internationalist activity was abetted by new technologies. Reliable railways and postal systems made communication swift and possible. So, too, did the 'tele' or distance technologies – the telegraph, the telephone and, even experimentally, telepathy, such as the kind investigated by the British Society for Psychical Research (which was also interested in Freud). After the First World

War, scientific internationalism also enjoyed a new wave of state and philanthropic funding.

It was in this context, at the urging of the younger generation of analysts – Sandor Ferenczi, Ernest Jones, Carl Jung, all talented thinkers and practitioners in their own right – that international meetings of psychoanalysts began, first in 1908 in Salzburg, then in 1910 in Nuremberg when an International Psychoanalytic Association (IPA) was formed under the presidency of Jung. By the time of the third Congress in Weimar, there were 106 members and two new American societies. Congresses were then held annually or bi-annually in a variety of capital cities. The IPA is now an international accrediting and regulatory body in the field, with some seventy constituent member associations that also oversee training institutes around the world. Psychoanalysis has also bred other kinds of schools and associations – Lacanian, existentialist, plus the many forms of psychotherapy – hardly uniform across the globe.

But I want to turn now to explore how Freud developed his own institutional linkages and networks after his early neurological and postdoctoral studies. Although not directly affiliated with or, you might say, not a tenured professor at the University of Vienna, Freud began to lecture as a *privatdozent* (a title granting permission to the qualified to lecture but denying them any power of backing from the university) on a fee basis in the 1880s. He became a 'Professor Extraordinarius' – a professor without a chair – in February 1902. During this time, Freud built up a number of different audiences, who helped disseminate his ideas: his professional colleagues – both in his personal correspondence network and the fellow members of societies – his peers, readers and students (both formal and informal). He gave public lectures and hosted seminars. He was by all accounts a lecturer of flair, speaking fluently and with an often casual directness – as the Clark Lectures he delivered in the United States in September 1909 readily reveal. Soon after he received the public affirmation of a professorship, Freud, in the autumn of 1902, initiated the 'Wednesday' evenings – a study circle (to begin with of a mere four attendees) at his home. These Wednesday meetings – with interested doctors, soon-to-be analysts, and visitors – where papers were read, discussed and hotly debated, soon became known as the Wednesday Psychological Society. Members were voted by the existing fellowship – not unlike the foundling Royal Society back in 1660, when science was still largely known as natural knowledge. The difference perhaps, though there were plenty of arguments at the Royal Society too, was that Freud's own, and indeed every country's, psychoanalytic society had a habit of splitting – a feature of movements as a whole, to judge by socialist or indeed religious groupings.

We should pause to ask whether psychoanalysis owes anything in particular to its mother tongue being German, and to its founder being a cultured and assimilated German-speaking Jew? This is a question with many possible responses, but it's important to remember that the initial reception of psychoanalysis was among those who could read German, if not necessarily Germans themselves. It's worth bearing in mind, too, that at the point in intellectual and scientific history when Freud founded psychoanalysis, German was an important global language, perhaps not as dominant as English became in the last third of the twentieth century, but a necessity for any intellectual or scientist of the period 1880–1920. German writing was thus international writing. We can speak of a specific reception of psychoanalysis in Germany and German-speaking countries, but, as it turned out, the future of psychoanalysis would not depend on that reception. In that sense, the reception of psychoanalysis in Germany was not privileged. There were inevitable conflicts and arguments, well documented in the early correspondences of Freud, Jung, Karl Abraham, Ferenczi and others. The German psychiatric and medical institutions, notwithstanding Jung's best efforts, were hostile

to psychoanalysis. Once Nazism took hold, despite Jung's continued adherence to Jew-free societies, the German movement
disappeared. It did not surface again until well after the Second
World War.

Freud had determined very early that the fundamental strategy and the one that would make psychoanalysis distinctive
was a turn to the people, a wider cultural community, and to
patients, rather than to the professionals. He wrote to Abraham
in 1907: 'You know yourself of the hostility I still have to con-
tend with in Germany. I hope you will not even attempt to gain
the favour of your new colleagues . . . but rather turn directly
to the public.'⁴ Patients themselves become a crucial audience
for Freud and an important means of circulation for psychoanalysis. This is true in some sense for all contemporary human
sciences – the objects of knowledge are also its subjects – but it
is perhaps particularly the case in the discipline of psychology
where students are expected to participate as experimental
subjects in professors' studies. In psychoanalysis the participa-
tion of the patient or analysand in the psychoanalytic process is
so extensive that it looks very much like collaboration. Indeed,
Freud called his early patients, his teachers (and you'll remember that the term 'the talking cure' came from a patient, Bertha
Pappenheim or Anna O). So, we should not underestimate the
importance of patients as agents in both the formation and the
dissemination of psychoanalysis. Remember too, undergoing
an analysis is a fundamental part of training to be an analyst.

One of the peculiarities in all this is, however, that psychoanalytic treatment is confidential. It takes place in private. Yet
despite confidentiality, influence spreads. An analogy might
lie in reading a book – a private act; but you might then give a
copy of the book to a friend or recommend the author. A 'taste'
network forms. Psychoanalysis produced a private network of
considerable power. This worked to Freud's advantage. Freud
was made a corresponding member of the Royal Society in the
1930s; he was nominated by the geophysicist Harold Jeffreys

who, unbeknownst to Freud, had been a patient of Ernest Jones in the 1910s. Many analysands had cultural significance and importance: for instance, the Governor of the Bank of England from 1920 to 1944, Montagu Norman, had psychoanalytic treatment in the 1930s. What effect might this have had on British economic policy or on the dissemination of psycho-analysis through the elites of Britain? We'll never have the full picture, but I suspect that this confidential network of patients, many of whom at least at first were members of the social and culture elites, played a not inconsiderable role in the growth of the psychoanalytic movement.

The wide circulation of psychoanalysis also owed something to its particular characteristics. One can never underestimate the influence of gossip, word-of-mouth and rumour, and many psychoanalytic ideas spread through these ordinary mechanisms. Psychoanalytic knowledge has also spread in interesting ways. Many people know things about psycho-analysis, even though the sources for their knowledge can't be traced. Freudian concepts and symbols, probably because of their shocking association with sex in an era that did not speak readily in public about matters sexual, quickly became dis-persed in the vernacular. People might assert, for example, that a snake or a cigar was a phallic symbol without knowing where they picked up this idea, and then perhaps giggle or snigger. It's an 'everybody knows' sort of knowledge: Freudian, often pseudosexual, knowledge deeply penetrated the twentieth-century cultural unconscious. In fact, in some cases, there is no traceable source for these kinds of symbolic associations. The well-known anti-phallic and anti-symbolization comment, 'sometimes a cigar is just a cigar', attributed to the many-cigars-a-day Freud, has spread as rumour: it has no textual origin. So too, the attribution of the term 'vaginal orgasm' to Freud, a phrase he never used in his writings. Such tantalizing and seemingly common-sensical ideas, abetted by the media, circulate wonderfully; like jokes, their origins are unknown but nonetheless cemented in consciousness through repetition. A proportion of psychoanalysis enters culture through this joke formula.

The concrete dissemination of psychoanalysis also takes place through influential 'seedbed' groups – writers, avant-garde artists and intellectuals, magazine editors, chattering Bohemians, café circles. Indeed, the spread of psychoanalysis, not unlike the modernism of which it is a part, can be tracked to certain intellectual cafés and artistic cabarets in Munich, Vienna and Budapest in the 1900s. By 1915, it was being discussed in the salons of Greenwich Village. In London, it found a home in the Bloomsbury-based 1917 club, initially an anti-war grouping of intellectuals and bohemians, famously including Virginia Woolf and figures such as the economist John Maynard Keynes and Ramsay McDonald, who would become the first Labour Prime Minister. These avant-garde intellectuals became the recipients and enthusiasts of psychoanalysis throughout the 1910s and '20s. This is a group who, of course, write, contribute to or edit magazines. In the writing of this time, psychological discourse moves into the mainstream, while psychoanalysis and various forms of therapy play their (often-enough negative) part in novels from May Sinclair and Virginia Woolf to Scott Fitzgerald and Italo Svevo.

The last group crucial to the dissemination of psychoanalysis – particularly psychoanalysis as a treatment – are the professionals: doctors, psychiatrists, psychologists, neurologists. These different professional groups relate to psychoanalysis in distinctive ways. Historically, during the late nineteenth and early twentieth century, the relationship between neurology and psychiatry was a complicated and adversarial one, both in Europe as well as in the United States, though with different emphases. The social historian of psychiatry Edward Shorter has offered a convincing and clarifying hypothesis for the reasons underlying the combativeness: neurology is urban, specialized, middle class.[5] In Britain, for example, it was centred in the National Hospital in Queen Square in the middle of London: this is still a hub of cutting-edge research. Psychiatry, by contrast, was back then consigned to the lunatic asylums (the mental hospitals, as they came to be called). Located on the outskirts of towns or in the countryside, these old state or charity-run institutions held a very low status in the medical profession, as for a time did those who practised within them. The clientele belonged largely to the lower classes, and most of the doctors lived in – just the other side of the wards' closed doors. A not insubstantial proportion of the asylum population in the late nineteenth and early twentieth century suffered from a range of geriatric dementias, acute alcoholism or tertiary syphilis (until a cure was found in 1909–10).

Shorter suggests that psychoanalysis (particularly in the US) provides a bridge between psychiatry and neurology, by taking psychiatry out of the asylum. This bridging characteristic facilitated psychoanalysis's dissemination. Psychiatry, was, on the whole, state-run caretaker medicine; in the US, asylums were state rather than federal institutions, and conditions were dire. (A confusion arises in the US in that the word 'psychiatrist' came also to cover psychoanalytic practitioners, once psychoanalysis became part of the training in psychiatry doctors receive through medical schools, a factor itself important for the institutional spread of Freud's ideas.) But in contrast, psychoanalysis offers up 'office psychiatry'. It takes place in consulting rooms in the centres of towns. It also takes place in elite clinics in attractive countryside, a fact that in turn unsettles the class associations of psychiatry. Take the example of Jung's practice. Like so many other psychiatrists at that time, Jung started his career as a live-in medic, in his case in the famous Burgholzli teaching hospital attached to the University of Zurich. Because of its growing fame for developing new psychological treatments, the hospital also attracted middle-class patients as well as visiting doctors from across Europe and the US. Freud's work was utilized by its head, Eugen Bleuler, an early reviewer of *Studies on Hysteria*, who was interested in research on the unconscious and its effect on mental life, including psychosis, before his junior, Jung, took it up. Jung, as a psychoanalyst, moved from the Burgholzli into private practice.[6]

Psychoanalysis bridges one gap in the provision of treatment, by making a kind of psychiatric treatment available outside the asylum for a low capital cost. In Britain, psychiatric care was given only to those who had been certified under the mental health act. Throughout the 1910s and '20s, both professionals and patients protested against this legal barrier to accessing psychiatric services. The barrier had some justifiable historical basis: the law could serve to protect against easy or simply punitive incarceration in an asylum, say by the sick person's family. It also enabled the use of asylums for those who found independent life impossible and who were indeed markedly more incapacitated than sufferers who might feel in need of psychological help. Psychoanalysis provided another area of treatment and one with a little less stigma attached.

Throughout the 1920s and '30s, a growing market for therapy of all kinds came into being – not only for the psychotherapies, but for the related developing disciplines of paediatrics and sexology. Practitioners in these new areas were often outsiders – those excluded from established disciplines because of their gender, ethnicity, religion or class. I once asked a colleague why there were so many Jews in theoretical physics. He replied: 'Well, they were the ones who weren't allowed to work in labs.' I leave it to you to decide whether that's a sufficient explanation, but it's undoubtedly true that at the beginning of the twentieth century certain subdisciplines – including psychoanalysis and paediatrics – come into being and 'outsiders', including women, play a significant part in their growth.' Psychoanalysis plays an important role in these new forms of knowledge and practice.

The last group of professionals who have a prominent role in the spread of psychoanalytic ideas and practices are those figures who emerge with what we might describe, following Michel Foucault's distinctive historical account of the development of the human sciences, as the rise of 'disciplinary' disciplines: probation officers, criminologists, psychotherapists, social workers. These new figures are born alongside the established professions such as medicine. Psychoanalysis certainly infiltrated ways of thinking in these new disciplinary professions. It also provided them, sometimes in a slippage from the law, with a distinctive idea of cases and a new version of individuality.

The First World War and its aftermath

In thinking about the history of psychoanalysis, it's worth pausing to consider the role of the First World War. The history of psychology broadly understood sees a number of crucial changes in these years. A key one is the introduction of the psychological category of 'shell shock'. This term was coined to describe a new large-scale, incapacitating, war-triggered condition with diverse symptoms – from amnesia, headaches, dizziness, tremors, tinnitus and hypersensitivity to noise, to conversion symptoms such as fugue and mutism – all linked to no evident brain injury or head wounds. The term was put into scientific circulation in England in *The Lancet* in a 1915 article by C. S. Myers. (Freud called the phenomenon the 'war neuroses', and gave the condition a differently nuanced aetiology and interpretation, itself discussed by the Austrian War Ministry, but the symptoms themselves crossed borders.)

Myers was both a doctor and an experimental psychologist at the University of Cambridge. Wartime saw him become an important senior figure in the Royal Army Medical Corps

(RAMC). He spent time on the front line and with men in the trenches. Shell shock was rife. Myers later regretted using the inexact but evocative term. Many of the sufferers (who would, after the Vietnam War, be diagnosed with PTSD) had not been close to shell fire, but the name did give a medical reality to conditions brought on by war that had earlier been judged as mere shirking and shamming behaviour. Shell shock has lived on as the primary term for describing what happened to hundreds of thousands of soldiers during the war. At the Battle of the Somme, some 40 per cent of casualties were diagnosed with it. A large cohort of doctors working for the RAMC were drafted in from other specialties to work with the shell-shocked soldiers. These doctors developed a psychological approach – sometimes more, sometimes far less based on psychoanalysis. However, towards the end of the war, it was widely accepted that a psychological approach was necessary for the treatment of such conditions.

The word 'psychological' can, needless to say, cover a range of hardly 'savoury' treatments – including electro-shock, one of the treatments for shell shock. This early form of the practice was made famous in Pat Barker's *Regeneration Trilogy* (1991–5), a series of novels about the twentieth-century physician and Cambridge ethnographer W. H. R. Rivers.[7] Electro-shock was used in the treatment of hysterically aphonic patients, those unable to speak; it involved placing electrical implements on the tongue or larynx of the shell-shocked soldier and turning up the electrical current until he spoke. This sounds – and was – cruel. It was torture. But it was psychological, not physical, torture. For it was an early psychological treatment; the electricity itself does nothing apart from transmit a relation of authority from the doctor to the patient. As in hypnosis, the rationale is to render the patient helpless, so that orders will then be followed, including the order to return to the Front. Arguably, this form of psychological torture was at least slightly less bestial than a punishment of death by firing squad.

What the place of psychoanalysis would be in the new medical and cultural landscape produced by war wasn't always clear. On the one hand, in the chauvinistic aftermath of war, there was a risk of 'imported' psychoanalysis being branded as a barbarian, Teutonic mode of thought (as it was regarded to a large extent in France), or, indeed, as a tainted Jewish perversion of an Aryan science. As Einstein notably quipped in 1919:

> By an application of the theory of relativity to the tastes of readers, today in Germany I am called a German man of science and in England I am represented as a Swiss Jew. If I come to be regarded as a *bête noire*, the descriptions will be reversed and I shall become a Swiss Jew for the Germans and a German man of science for the English![8]

The post-war period witnessed this swirling range of nationalist and geopolitical forces in which the new cultural and scientific movements were caught up. On the other hand, during the 1920s, there was considerable enthusiasm for psychoanalysis, thanks to its implicit attack on authority. This seems consonant with the attack on the world of the 'fathers', those leaders who had tumbled nations into the killing fields of war. Psychoanalysis took on the aura of a great panacea. For the bestselling writer of the period, H. G. Wells, in a piece titled 'The gifts of the new sciences' in *The Strand Magazine* of February 1924, 'what is of the greatest promise in the science of the present time is the new study of human motives which centres around psycho-analysis'. He continued:

> The past century has been the supreme century of material achievement; the present and the twenty-first centuries will, I believe, be the great fruiting and harvesting time of psychological and physiological science. Man having run all over his world from pole to pole, having learnt how to fly round it in

seven or eight days and how to look or speak round it in a
flash, will presently, I think, become introspective and turn his
practical attention to himself.

Wells predicted that the next hundred years would be
'essentially a century of applied psychology', during which
a 'new revolution' based on psychoanalysis would bring
about changes in art and literature and produce 'an increasing tendency to psychologize legal, political, financial and
economic conditions'. Surprisingly perhaps, J. D. Bernal, the
great physicist and mathematical crystallographer who was
writing during this post-war period about Einstein's theory
of relativity, seemed to agree with Wells. Rather remarkably,
Bernal states that the great scientific revolution of the day is
psychoanalysis (not relativity theory which he treats as just an
interesting exercise in physics).[9]

So what is this movement that is so enthusiastically sup-
ported in the early 1920s? There is a utopian dimension to
psychoanalysis during this period. Not only has it transformed the relationship between doctor and patient – a
transformation that will work itself out in many different
and unpredictable ways across the twentieth century. But
psychoanalysis is also now seen as part of the march of
truth. It emerges as a sequel to the great Enlightenment project in which truth acts as a weapon to be wielded against
superstition and oppression. Freud's work, perhaps not unjus-
tifiably given his sociocultural focus in these years, takes on
the aura of a revolutionary force. Psychoanalysis is likened to
Bolshevism: it is an attack on the oppression of contemporary
institutions and their power, represented by the father, an
out-of-date patriarchal authority who led the world into war.
In this post-war moment, Freud's account of the Oedipus
complex is read as a radical undermining of patriarchal society. (This interpretation of the Oedipus complex will shift
and later be reversed by 1970s feminist critics, who will be

pointing out that Freud's edifice is itself patriarchal and oppressive.)

The emerging psychoanalytic movement can also be understood alongside the other comparable political and health movements that emerge in this period. For instance, the moral hygiene movement, which has some overlaps with psychoanalysis in its attempt to prevent mental illness, was conceived as a way of creating good habits in society. A public-led movement for psychiatric reform, it garnered the support of many doctors (and psychoanalytic insights were imported into that movement in a range of ways). Principally, however, psychoanalysis should be understood as part of the movements for sexual reform, part and parcel of the modernist thrust of the 1920s. The First World War had given many, most vociferously the young – artists, social and political revolutionaries – not least in Russia, but also across Europe, the desire to throw off the past with its burden of a history that had led to the war's barbarism. The first waves of the feminist and gay movements mounted attacks on conservative ideas. 'Free love' became a revolutionary slogan. The institutions of family and marriage, not least because so many men had been killed in the war, were in crisis. Longer-term shifts were also occurring in social structures and attendant values, particularly among the middle classes. Divorce became more common. Servants were increasingly difficult to find; middle-class mothers, either widowed or with maimed husbands, could no longer afford nursemaids, and so were thrust into having to have direct daily relations with their small children. (The rise of paediatrics is not unlinked to this.) Psychoanalysis became a tool in these wider social transformations.

National cultures of psychoanalysis

Psychoanalysis developed distinct national cultures early on. Take the reception of Freudian ideas and practice in the US. After Freud delivered his influential introduction to psycho-analysis, the Clark University lectures in 1909, psychoanalytic thinking penetrated many aspects of psychiatry and even the discussions of insanity in the courts. Unlike in Europe, Freud and his views on sexuality also moved very swiftly into popular consciousness – through glossy magazines, the press and the movie industry. In the 1930s, with the influx of European émi-grés in flight from Hitler, a far-reaching psychoanalytic culture developed and flowed into medical school teaching. In 1946, the Menninger Clinic in Topeka, Kansas, already famous for its work with children, opened a School of Psychiatry, heav-ily influenced by psychoanalysis and psychodynamic practice. This approach became the norm throughout the profession in the post-war years, when the needs of returning soldiers were to the fore. Erik Erikson's ego psychology and Heinz Kohut's self-psychology became two prominent strands within American practice. The influence of psychoanalysis within psychiatry lasted at least until the early 1970s, when drug treat-ments began to displace its importance. (Some would say the psychoanalytic moment has now once more returned, if again in an altered guise.)

Some of the triumphs of American films can be seen as the products of this psychoanalytic culture. Hollywood's early love affair with psychoanalysis produced a number of well-known films and further popularized treatment, some underlying para-Freudian ideas and the profession itself. Most famous is undoubtedly Hitchcock's *Spellbound*, about an amnesiac psy-chiatrist (Gregory Peck) who, with the help of a loving colleague (Ingrid Bergman), has to discover a traumatic death in his own childhood in order to release himself from the possibility of having murdered the colleague whose name he bears. American psychiatrists had played an important role during the Second World War – in marshalling morale, in treating new versions of war neurosis and indeed in providing character profiles of the enemy, especially Hitler. Hollywood versions of these later war neuroses often included a retrieval of memory from deep amnesia. In 1962 came John Huston's *Freud*, which put the man himself on the big screen, played by a nervy Montgomery Clift. This biopic gave a historical account of the early days of psychoanalysis. Huston had initially asked Jean-Paul Sartre to write the screenplay with Marilyn Monroe in mind to play a composite hysteric made up of Anna O and other early cases. Sartre's synopsis, at ninety-five pages, was the length of an entire script. Huston ordered a cut version, but what Sartre offered would have come in at eight and a half hours. Huston fired Sartre, cut up his script, and eventually made a very good film – a film that is both a biographical study of Freud and a historical account of the origins of psychoanalysis.

By the 1960s, Alfred Kadushin, a sociologist conducting a statistical study of psychiatry, would write of New York that 'over 70 per cent of the culturally sophisticated – that is, those who were likely to go to concerts, cocktail parties, plays, and museums or art galleries at least several times a year – were also members of the Friends and Supporters of Psychotherapy'. He is not referring here to an actual society but to a loose cultural alliance of therapeutic supporters and sometime patients. New York was, in this instance, replicated in Hollywood. At the same time, psychoanalysis, outside of the local and regional Institutes – there are now thirty-four training institutes affiliated to the American Psychoanalytic Association – was being taught within the medical specialization of psychiatry. Freud had hoped for a lay profession, alongside but separate from a medical psychiatric one, but the American organization of psychoanalysis precluded this. This meant that the non-medical analysts who fled Nazism to arrive in the US could not practise within the existing psychoanalytic societies, which demanded medical degrees of their candidates. (Today, those with graduate degrees in psychology and other mental health disciplines are also permitted entry to a postdoctoral psychoanalytic training.)

In France, psychoanalysis had a limited influence before the Second World War, though here, too, certainly at the start, there was a distinct division between cultural dissemination and the spread of professional practice. One important early path of entry was artistic, through the Surrealists and André Breton, who travelled to see Freud. A second literary route is associated with the writer and later Nobel winner André Gide, who had had access to Freud's theories through his summer visits to Cambridge with the Stracheys during the First World War. He stayed in Merton House on Queens Road, drank a great deal of good wine from Trinity College, and learned about Freud alongside his radical and often pacifist Bloomsbury Group friends. In his 1925 novel *The Counterfeiters*, a key character is based on the Polish analyst Eugénie Sokolnicka.[10] Sokolnicka had been analysed by Freud and was one of his emissaries to France. A brilliant woman, she failed to prosper after initial success with her practice and she committed suicide in 1934.

Meanwhile, the post-war alliance between surrealism and psychoanalysis burgeoned. Indeed, Surrealism, with its focus on dreams and a different kind of free association, initially thinks of itself as a psychoanalytic enterprise: Freud is their patron saint. By contrast with this literary and countercultural world, the first practising French analysts – apart from those who came from Germany or, in the case of Eugénie Sokolnicka, from Poland – have a Catholic leaning and are suspicious of the importance Freud accords to sexuality. They are also occasionally anti-Semitic. Indeed, their Catholicism is such that one of them, Maryse Choisy, who had very briefly been in analysis with Freud, seeks approval for psychoanalysis from Pope Pius XII. The Pontiff is prepared to approve of psychotherapy, but not of Freudian analysis: 'It is not proven, it is even untrue that the pansexual method of a certain school of psychoanalysis is an integral and indispensable part of any serious psychotherapy worthy of the name.' Sexual causes, the Pope told Choisy's group, violate the confession.[11] Only the redoubtable Princess Marie Bonaparte, Freud's analysand and emissary to France in the later twenties, has the stamina and skill, at least initially, to propel the Paris Psychoanalytic Society along more or less Freudian lines.

The most important figure to come out of French psycho-analysis, Jacques Lacan (1901–81), is widely known as the French Freud. Lacan, too, sought a kind of papal blessing for himself and his school in the 1950s, though less perhaps on religious grounds than in order to establish the importance of his then new dissenting school. Lacan undertakes his medical and psychiatric training in the 1920s and '30s, first with the liberal figure of Henri Claude, who gave his trainees the free-dom to experiment with Freudian ideas; then with the famous Gaëtan Gatian de Clérambault, the brilliant if eccentric clini-cian who described erotomania. At the same time, Lacan is completely bathed in the psychoanalytic culture of surrealism. A friend of André Breton's, he invites the group around the surrealist magazine *Le Minotaure* to decorate the walls of the Saint-Anne Hospital where he works. He contributes to the magazine side by side with Dali on the subject of paranoia, which is also the core of his thesis.

Before the war, Lacan had already begun to elaborate the dense conceptual apparatus that characterizes his work together with a marked fidelity to Freud.[12] His famous concept of the mirror stage begins then, as he develops an interest in the observation of babies and animals. In Lacan's early for-mulation, the developing child in front of the mirror suddenly recognizes herself for the first time, herself as other; through a process of identification, the ego is formed. Here, Lacan draws on Freud's understanding of narcissism and marries it to the

Hegelian dialectic of master and slave. For Lacan, there is a misrecognition at play here: the moment when the six-month-old sees itself as an enticing unfragmented whole in the mirror is undoubtedly prompted by parents speaking 'the discourse (and desire) of the other' ('What a beautiful girl', etc.). Thus, the ego emerges as an alienated object, selfhood as an illusion covering over the child's continuing helplessness.

Through his engagement with structuralist anthropology, linguistics and cybernetics, Lacan developed an account of language as constitutive of the human and of desire, and a theory of language as constitutive both of the unconscious and the Symbolic Order. Lacan will eventually replace Freud's triadic intrapsychic structure of id/ego/superego with his own triadic system of Real/Imaginary/Symbolic. In shorthand, the Real is the pre-linguistic, what cannot be symbolized or made conscious; the Imaginary is the internalized image of the ideal, whole, unfragmented self – akin to Freud's ego, which mediates between the inner and the outer world. The Symbolic is the world of language, of received forms and ideas into which the child grows, a kind of cultural superego.

Like Freud's, many of Lacan's ideas have moved into the broader French culture (as well as into various others, particularly in Latin America): Lacanian thoughts – the idea, for instance, that every demand that is not fulfilling a simple need is a demand for love, while desire can never be fulfilled and feeds on (dis)satisfaction – are, like, for instance, Freud's, 'uncanny', a part of the contemporary literature armoury. But Lacan's many innovations – his notion of *jouissance*, that is, the enjoyment that is 'beyond the pleasure principle', his important claim that the unconscious is structured like a language, his '*objet petit a*', that unattainable/fantasized object of desire – are beyond the scope of this lecture. Suffice it to say here, it was not these ideas that caused the 1953 split in the Société psychanalytique de Paris (SPP). He was then the president of the Society, initially set up with Freud's followers, not least

Marie Bonaparte (nick-named, 'Freud-a-dit'), and Lacan's own analyst Rudolph Loewenstein. Lacan argued the Society was no longer faithful to Freud. Things came to a head over Lacan's thinking on the nature of time in relation to action and hesitation, which led him to vary the length of analytic sessions. Many analysts saw this as heresy. In 1963 the International Psychoanalytic Association cited it as the reason for terminating Lacan's status as a training analyst (not a practising one). This was the basis of what Lacan described as his 'excommunication' from the IPA. It led to the collapse in 1964 of his new society, the Société française de psychanalyse (SFP) which had been set up in 1953, since the IPA insisted that the SFP would only be reintegrated into the IPA on the condition that Lacan be prevented from contributing to the psychoanalytic training of new candidates.

But the ten-year life of the new society around Lacan – and the École Freudienne de Paris that was set up after its dissolution – meant that a Parisian culture of radical psychoanalysis developed very differently from that of New York or London or Berlin. Beginning in the early 1950s, Lacan had held weekly seminars at the Hôpital Saint-Anne for the SFP, as part of their training programme. But Lacan's ideas of what psychoanalytic training consisted in, entailed, as Freud's had in his day, on drawing on the whole gamut of challenging thought in his historical moment – ideas from philosophy, linguistics, anthropology, the human sciences. Lacan refused the more traditional boundaries of medicine or the brain sciences. In his wake, psychoanalysis in France had alliances with philosophy and the human sciences more than with medicine and psychiatry, as was the case in the US. When he was barred from training analysts in 1963, his seminars remained remarkably popular and grew ever larger, moving over the years from the Hôpital Saint-Anne to the École normale supérieure (rue d'Ulm) and then to the Faculté de droit of the Sorbonne. By the 1970s, audiences numbered more than 1,000 and had long included the cream of French intellectual life, from Julia Kristeva to Louis Althusser, and many more. Through these seminars and their gradual diffusion, Lacan was famous not only in France but across the Western world of ideas. When he died in September 1981, the event was featured with full obituary roundups on cinema screens as well as television across France.

The Lacan effect has meant that French psychoanalysis was fundamentally intellectual before it was professional, linked to the expansion of the universities that was taking place worldwide. One could say that Lacan was the first analyst since Freud to mobilize the entirety of his culture's resources in building his analytic edifice.

Psychoanalysis in Britain

British psychoanalytic culture was a different kind of beast. Psychoanalysis initially entered Britain along multiple routes between 1895 and 1925. One of the earliest encounters in the UK was through the Society for Psychical Research in Cambridge, a group set up to scientifically investigate psychological events. These now seem both weird and wonderful with their focus on spirits and ghosts and telepathy. But the Society also reported on Freud's work as early as 1893.[13] The sexual reformer, Havelock Ellis, who trained but never practised as a doctor, wrote ground-breaking studies of homosexuality and the most encyclopaedic study of sexuality in English before the US Kinsey reports in the 1940s. Banned and on occasion prosecuted, Ellis was nonetheless widely read in radical circles. His assessment of Freud was favourable, as was Freud's of him.

Amongst the key early proponents of psychological developments within psychiatry was the brilliant Cambridge anthropologist W. H. R. Rivers. In London the medic Bernard Hart, who had studied psychiatry in Paris and Zurich and had worked in asylums, became the first practitioner in psychological medicine at University College Hospital, as well as an early 'shell shock' doctor, working at the well-known military hospital in Maghull. In 1910, Hart had written an influential paper on the unconscious in Janet and Freud. In 1913, an important year for psychoanalysis in Britain, Ernest Jones, effectively Freud's London ambassador, set up the London (soon to be British) Psycho-Analytical Society; A. A. Brill's translation of *The Interpretation of Dreams*, its first, was published and substantively reviewed; and the Medico-Psychological Clinic was established with the aim of treating anyone in need, therapeutically. Soon renamed the Brunswick Square Clinic, its early staff included notable feminists, as well as the modernist novelist May Sinclair. It became a well-known centre for the treatment of war neuroses. With a slight altera-tion of circumstance, Virginia Woolf's shell-shocked Septimus Warren Smith, in some ways Clarissa Dalloway's tormented, hallucinating double, could have ended up in Brunswick Square seeking help.

One important route was the Bloomsbury Group, a key disseminator of psychoanalytic interest in Britain. Not only did Leonard Woolf publish Freud at the Hogarth Press, John Maynard Keynes review him (anonymously), but four of the group's members, including Virginia Woolf's brother Adrian, trained as analysts, while James and Alix Strachey travelled to Vienna where they were both analysed by Freud. (Alix went on to have a second analysis with Karl Abraham in Berlin.) They both became Freud's translators, while James went on to edit *The Standard Edition*.

Freud's work in Britain was also taken up by a strand of progressive educators, including A. S. Neill, who founded the famous 'free school', Summerhill. Bertrand Russell gave up his career as a philosopher to become, for a while, a school-teacher, founding the Beacon Hill School with his then wife, the feminist Dora Black. A differently emphasized strand of progressive education, with links both to Freud and the Swiss psychologist Piaget, was key to the setting up of the Cambridge Malting House School. Perhaps most surprising in the British admixture was the excitement about psychoanalysis amongst scientists – not primarily those in the biological sciences or the medical sciences, but rather the physical sciences. Amongst them were Lionel Penrose, Adrian Stephen, Frank Ramsey, J. D. Bernal and Harold Jeffreys.

Britain's early clinical culture was very specialist, heavily focused in London, and made up of eccentric clinicians and theorists. With an elite model of organization and influence, it revolved around a disaffected, educated bourgeoisie – as well as some émigrés: Melanie Klein in the mid-1920s and then others in the 1930s including, by 1938, Freud and his daughter Anna in flight from the Nazis. After the Second World War, psychoanalysis, already heavily centred on work with children, begins to form very strong links with the NHS. The paediatrician and analyst Donald Winnicott famously presents BBC broadcasts on mothers and children, while John Bowlby and his attachment theory are key to the Tavistock clinic. Through the Hungarian analyst Michael Balint, and his important book *The Doctor, His Patient, and The Illness* (1957), psychoanalysis provides a key set of ideas for – even the foundational ideology of – the general practitioner in the NHS. This extremely influential book on the use of psychoanalysis in general practice is still cited on the Royal College of General Practitioners website as the founding text for what it means to be a GP in the NHS.

A distinctively British strand of psychoanalysis develops from the 1930s on as a number of influential psychoanalytic thinkers diverge from Freud's emphasis on instincts to articulate an account of 'object relations' – an analysis of the complex relations of defence and desire that the subject takes up in their internal relations, as well as their relation to the world of 'objects'.[14] It's important to remind ourselves, that by the mid-twenties Freud's theory of the instincts or drives was no longer, if it had ever been, altogether biological. The instincts (first the sexual, pleasure-seeking, libidinous and self-preservative, then the death instinct, which separates, destroys, brings conflict and the compulsion to repeat) were 'a concept on the frontier between the mental and the somatic, as the psychical representative of the stimuli originating from within the organism and reaching the mind, as a measure of the demand made upon the mind for work in consequence of its connection with the body' (SE XIV: 121–122). As the ego mediates between the demands of reality, the instinctual pressures from the id and the moral prohibitions of the superego (amplified by the journey through the Oedipus complex and the massive injuries to narcissism associated with that), object relations, relations to those internalized or on the outside of the subject, both desired and defended against, take on increasing importance. Central to several of the leading British analysts is the psychoanalysis of children.

A key figure in the analysis of children is Freud's daughter Anna (1895–1981), his 'Antigone', at once his ambassador in his later years and the guardian of the Freud flame. She writes *The Ego and the Mechanisms of Defence* (1936) two years before arriving with the family in London where, during the war, with her partner Dorothy Burlingham, she sets up the famous Hampstead Nurseries. In this book, which some have called a textbook – one, however, written with exemplary clarity – she explores how the ego protects itself against instinctual demands. She is interested in the psychology of the defences.

At the centre of Anna Freud's conception of psychoanalysis is the importance of strengthening the ego, so that it doesn't have to resort to primitive or pathological defences. 'From the beginning', she writes, 'analysis, as a therapeutic method, was concerned with the ego and its aberrations: the investigation of the id and of its mode of operation was always only a means to an end . . . the restoration of the ego to its integrity.' Anna Freud's orientation, ever pragmatic rather than theoretical, provides the basis for what becomes ego psychology in the United

States: it's a psychology that believes in the normal. Particularly astute in her insights into teenagers – their pendulum swings between asceticism and self-gratification, a re-enactment of earlier infantile sexual configurations – she gives an ever-lucid account of ten ways in which the ego defends itself against the incursions of the id: repression, reaction-formation, projection, introjection, regression, sublimation, isolation, undoing, reversal and turning against the self. To these she adds two novel defence mechanisms, which turn on innovations within the concept of identification and amendments to the Oedipus complex, each stemming from her own analysis: identification with the aggressor and altruistic surrender. You'll perhaps be familiar with the first from its political examples, say in Frantz Fanon's *Black Skin, White Masks*, in which the oppressed subject internalizes the values of the colonizer and judges himself by them to his own detriment. The second new defensive syndrome, altruistic surrender, is exemplified in the case of a governess who lives wholly through others, because an early 'narcissistic mortification' has seen her displace her wishes onto those better qualified to fulfil them.

Anna Freud herself – who, after the Nazis invaded Austria, always carried veronal with her – acted as a bulwark in protection of her family. She had long been Freud's ambassador and, with his gathering illness, she became his deputy and a central figure in the international organization of psychoanalysis. In her analytic work with children, Anna perhaps never altogether forgot that she had trained as a teacher: she always thought of the analyst as working in cooperation with parents and educators: babies and small children were not yet fully formed. Her work had a determining influence in shaping child and family law in both the US and the UK: we owe her the term and the rulings 'in the best interests of the child'.

Anna Freud was preceded in the UK by another émigré, another child analyst, the Vienna-born Melanie Klein (1882–1960) who arrived in 1926 at the invitation of Ernest Jones and was already known to several members of the British Society. A forceful woman, both in manners and intellect, she had been analysed by Freud's friend Ferenczi in Budapest and Karl Abraham in Berlin. The battles between the two women and their groups of followers were never altogether resolved and were at their height during the so-called 'Controversial Discussions' of the war years when methods of training were at issue. Given the mother-centredness of analysis in Britain, it's clear that Klein's innovations had a greater influence on both theory and practice (Klein was also influential in Europe and South America, though her work had a hostile reception in the US). Klein's innovations are numerous. She stressed that since humans relate to each other from the earliest moment, transference is always, however young the child, active This meant that child and adult analysis can follow the same principles: the key difference is that the material in a child analysis is the language of play, rather than the language of voice. Klein became famous for her play analysis: she used toys and play in her work with young children, arguing, in contrast to Anna Freud's vision of the pedagogic function of child analysis, that these were the child's symbols and so the language to be interpreted when they brought their fears and anxieties into the consulting room. The here and now of the analysis itself (rather than any links with parents or educators) were the crucible of her work.

The mother is central to Klein's theorizing. Whereas Freud thought of the early period of an infant's life as pre-Oedipal, Klein does not allow for any Freudian pre-Oedipal moments: the mother is an Oedipal mother from the very first, with guilt and unconscious phantasy already at work in the baby at breast – a breast which the infant takes in as an object and which can be persecutory as well as idealized. Klein emphasized the 'internal world', with its 'internal objects', and replaced Freud's instinctual phases with a less chronological mode of relating to the internal world – she talks instead of 'positions'. Relations to the breast dominate the first six months of the child's life and constitute what Klein names the paranoid-schizoid position. Defences such as projection and introjection, splitting and projective identification, come into play: these are not unlike what an adult psychosis might entail. Once the object becomes a whole, with both good and bad parts, a more unified ego becomes possible – as does a state of mourning over the fact that the object now actually disappears. When the infant discovers her helplessness and loses her omnipotent object, she moves into the depressive position and ambivalence becomes key to her state. All this and more – from the inchoate emotions that attack the infant to envy and gratitude – are grist to Kleinian analysis.

The great Donald Winnicott (1897–1971), who was arguably Britain's most influential post-war analyst in his effect through his broadcasts on public life, was himself a practising paediatrician. Certainly, he is the analyst who in writing creates the most English of contributions to therapeutic culture. His model of psychoanalysis also had a mother–child orientation, more developmental in his case than in Klein's. For Winnicott, the ordinary 'good enough mother' laid down the basis of a child's good health in the earliest months of a child's life. To this, 'holding' was key. For Winnicott, at first, there is no baby, only a mother–child relationship. The mother must perform the function of containing the child, so as to let inner maturation become possible. Winnicott's account of child development also offers an emphasis on the 'transitional', on transitional objects and phenomena. These objects, the first 'not-me' objects – the teddy bear, the blanket – become the world of reality (in contrast to the 'internal objects' developing with the mother); their abandonment and destruction allows a capacity for growth.

What the analyst metaphorically enacts is this reliable holding where it may have failed the child: he 'holds' the patient, allowing the very presence of the analyst/mother to contain the destructive, aggressive forces the child fears. For Winnicott, an interpretation in analysis, if it is well-timed and correct, can give the child a sense of being held physically. Like Klein, it is through play that child-analysis unfolds. But this play holds an added significance for Winnicott, since it is through imaginative play that the child develops 'a capacity for being', the ability to feel genuinely alive, or real. If that fails, she develops a false self – a kind of mask that complies with the expectations or demands of others that have been internalized or introjected. For adults, play is equally reparative in the creation of an authentic self – and there are many kinds of play, from making art to making good conversation to having a good analysis.

From these three examples – Anna Freud, Klein and Winnicott – we can get a sense of how in Britain, more than anywhere else, the life of the child informed psychoanalytic thinking and practice far more than sexuality. But I don't have time here to compare this in detail to other national psychoanalytic cultures. For instance, the development of psychoanalysis in Latin America is a particularly complex story, crucially in Argentina and Brazil. There it was initially a badge of modernism – a kind of technological modernism of the self. It only grows later into a profession, in part, once more, with the arrival of European immigrants after the Second World War. In Argentina, psychoanalysis takes off in the 1960s, especially after 1968. Its reach, particularly in the 1980s, extends to include working-class communities, much more so than in much of Western Europe and the US. Like in Italy, psychoanalysis in Latin America is disseminated into popular culture through the mechanism of the 'agony aunt'. There are separate stories to be told, too, of Germany – a country that was pivotal to early analysis but where the arrival of the Nazis decimated the profession (Freud's books were among those burned). India and Japan, the Middle East and China, also have their own stories.

Before I finish this lecture, I want to make two points about the history of the psychoanalytic movement and to take a

moment to draw a different kind of map of psychoanalysis –
one that is far less tied to these dispersed national cultures with
psychoanalytic traditions of their own, and more in keeping
with the internationalist spirit of the movement I discussed
earlier.

There is a basic model of psychoanalysis that cuts across
national cultures: the model of how one becomes an analyst.
Early on, Freud believed that one became an analyst through
the interpretation of one's dreams (as he had done) and
through analysis; this was at the core of psychoanalytic training from the outset. A personal analysis was a prerequisite
– for Freud, but also for Jung, Eduard Hitschmann and other
early members of the profession. By the 1920s, however, it
became imperative to establish a more formalized training
model that included more than a training analysis. Such training allowed for a distinction between the 'qualified analyst' who
had learned the bases of psychoanalysis and the 'wild analysts',
to borrow the term coined by Freud in 1910 (SE XI: 219–227),
such as Otto Gross (who used sexuality – literally sex – as an
analytic 'tool'). Under the leadership of analysts Karl Abraham
and Max Eitingon, who also put up the funds, Berlin became
the first centre to establish a psychoanalytic training institute,
alongside its free 'Policlinic'. Vienna followed soon after in
1924, with Helena Deutsch as its coordinator and president
until 1935. The International Training Association established
a set of standards in 1925 (thereafter it became less a 'scientific society' than a site of professional training). The Training
Model included three key elements: personal analysis, supervised cases (in which the trainee analyst discusses their cases
with a supervisor) and seminars, the basis for clinical communication, which involved the presentation of clinical material
to peers and students. These remain in place today.

The system of training, consisting of analyses and supervi-
sion by a more senior analyst, created a complicated web of
connections and affiliations between analysts. 'National' ways

of thinking grew out of this web. Early on, however, while the first generation of analysts were still working, national webs were less important than the pervasive influence of the first group of analysts: many British, French and Americans travelled to Vienna and Berlin for analysis. This flow had its later counterpart in the German-language analysts fleeing continental Europe to settle in the UK and the Americas. Freud scholar Ernst Falzeder represents the supervisory analysis effect in his staggering 'spaghetti junction' diagram.[15] The map lays out a history of psychoanalysis that is also a social structure of the discipline. Freud is at its centre; each line extends outwards from him in a proliferation of links through analyses. The resulting web is like a complicated family tree, a genealogy. Magnifying the diagram, we see that a number of Freud's own family went into analysis, though mostly not with Freud himself, and went on to become significant figures in the history of psychoanalysis.

In the original form of this map as conceived by Falzeder, there are four different types of link: (1) analysed by, (2) supervised by, (3) has sex with, (4) marries. The presence of these intimate connections points to a danger within the profession. Since lying on the couch necessarily involves a level of personal vulnerability, things can sometimes turn messy in the atmosphere of psychoanalytic institutions. Though sexual connections are undoubtedly transgressive, they did often enough occur (perhaps more so in the past than now). Within this spaghetti network, then, there exists a tangle of professional, personal and family links. These give a distinctive cast to the psychoanalytic movement as a whole.

I want to close with a final thought, which connects this history of the psychoanalytic movement to our exploration of Freudian social theory in an earlier lecture.[16] Then, we were considering what conception of the social is proper to psychoanalytic theory – and we looked at how Freud scaled up his analysis of the individual to society and culture. But there is another conception of the social at the heart of psychoanalysis, one that gets disseminated through the psychoanalytic movement. This is the conception of the social built on the transference. One way to understand psychoanalysis is to see it as a guide to modern, alienated individuals for whom community and family ties become irremediably loosened. What the clinical practice of psychoanalysis teaches is that transference is at the core of all psychoanalytic relationships. Psychoanalysis is a Socratic discourse intended to free the patient's truth insofar as they have constituted that truth in the discourse they are engaged in with the analyst. (As Lacan had it, the unconscious is the discourse of the other: what the analyst says is the implicit meaning of what the patient has been saying.) This transferential relationship is engineered so that the relationship itself becomes the site of truth; it becomes the patient's symptom. Hence transference is not only the empirical finding of analysis, but its major tool, which is itself constructed by the techniques of analysis. In this sense, the transference should be the principle of social organization of the psychoanalytic movement, since in essence it is the principle of the social in general. If we take this view seriously, then we can see that psychoanalysis becomes social and universal *through* the system of training (with the transference at its core) and therefore through the history of the psychoanalytic movement itself.

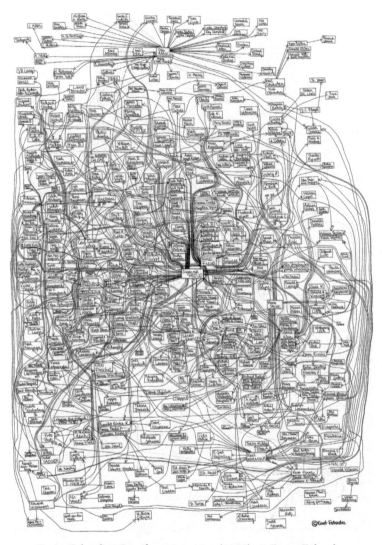

Ernst Falzeder, *Spaghetti Junction* (1993). © Ernst Falzeder

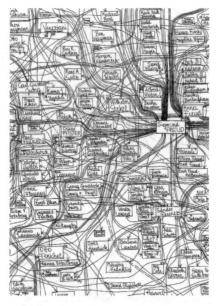

Close-up of *Spaghetti Junction* (1993). © Ernst Falzeder

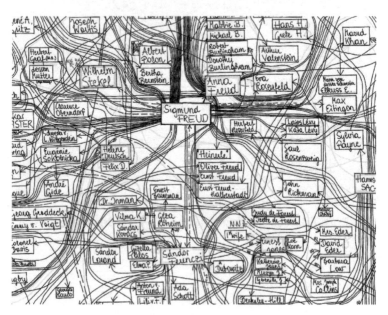

Close-up of *Spaghetti Junction* (1993). © Ernst Falzeder

Lecture 6

The Significance of Psychoanalysis in the Twentieth Century

(Left to right) J. D. Griffith Davies (Royal Society Assistant Secretary), Sir Albert Seward, Sigmund Freud and A. V. Hill at Freud's home in Hampstead, after Freud's admission to the Royal Society in 1938. In an extraordinary gesture, the Society's historic Roll Book was brought out of its headquarters to a frail Freud, then still at his first London address at 39 Elsworthy Road, for signature. Portrait of Sigmund Freud and others by Madame Marie Bonaparte, 1938 © The Royal Society

Arguments against Freud and psychoanalysis, whether scientific, historical, philosophical, ethical, political or common-sensical, have raged ever since the field Freud put on the intellectual map had a name.[1] Take the range of historical arguments that have shown the limitations of psychoanalysis. Freud's thinking stands accused of having recourse to an outdated biology. He has been portrayed as a scientist whose theories have little evidential base. Others, who see him as more of a social critic, stress his lack of originality, by showing how the concept of the unconscious has a variety of historical precedents that predate Freud's distinctive account of it.

The late culture theorist, Malcolm Bowie, in his essay 'Freud and the European unconscious', sums up these sorts of criticisms with panache:

'How original was Freud?' is still a question of seemingly inexhaustible historical interest. A characteristic sequence of answers and counter-answers to it might go like this: (1) he was not original, other than as a publicist, because the unconscious had come and gone in European thinking since antiquity and become positively fashionable during the period of Freud's early maturity; (2) but Freud differed crucially from his predecessors in that he was a thoroughgoing systematist, and promoted the unconscious only in so far as the psychodynamic system of which it was part could explain hitherto unexplained mental facts; (3) but the very notion of 'system' that Freud resorted to was a commonplace of the new and topical evolutionary biology that he grew up with, and even when he repudiated biological science in favour of a supposedly 'pure' psychology he was still adhering to a biologically inspired theoretical mode; (4) but in doing so he was exploiting biology for his own purposes, not remaining subservient to it: all spectacular paradigm-shifts in the history of science begin with a switching or mixing of metaphors; (5) but this is exactly the problem with Freud: he metaphorized science; (6) but . . . At each stage in this argumentative game as it is nowadays played, Freud's abstract mental models are re-immersed in their native sea of particulars, and his family history, education and social background – together with his professional relationships, reading habits and pastimes – are relentlessly trawled and dredged. Can any self-proclaimed universalist ever have been returned with such self-righteousness on the part of his commentators to the local habitat in which he and his ideas were born?[2]

In this way, psychoanalysis has been shown to be limited in historical time and space, with Freud's ideas tied to their Austro-Hungarian context and Freud himself bracketed as a Victorian moralist. Some have argued that Freud fabricated his cases or that, when judged by the scientific standards of his own time – as well as ours, thanks to shifts in the criteria of scientificity – he falls short.

Even by its advocates, the reception of psychoanalysis has often been ambivalent. The twentieth century is littered with great thinkers whose work bears the mark of Freud's influence, who have, at the same time, criticized his work. For the economist John Maynard Keynes, Freud's 'scientific imagination' generated theories of 'great and permanent significance' alongside those that 'will have to be discarded or altered out of recognition'.[3] For the philosopher Ludwig Wittgenstein, Freudian explanations were plausible, but their status was merely that of redescriptions. This ambivalent reception has also been reflected in the changing politics of psychoanalysis. In this final lecture, I want to talk about this politics – and to explore the cultural, scientific and political significance of Freud's thinking and of psychoanalysis in the twentieth century.

The progressive politics of psychoanalysis

Politically, psychoanalysis has long been understood to be a subversive force – if not always as a therapy, then certainly as a movement and as a set of ideas. Freud's view of the human subject is one that undermines the very possibility of authority, of certainties, of fully rational choice – and in that it is already fundamentally revolutionary. At the foundation of psychoanalysis is a criticism of sexual injustice, a subversion of a repressive system and of the moral codes of sexual life. Psychoanalysis makes what the culturally conservative like to consider as aberrant – particularly where sex is concerned – into ordinary-enough traits or behaviours. Freud's assertion in 1905 in his *Three Essays on the Theory of Sexuality* throws down a gauntlet to convention and conformity: 'The disposition to perversions of every kind is a general and fundamental human characteristic' (SE VII: 191). A little later, in the spirit of radical liberalism that is fundamental to his '"Civilized" sexual morality and modern nervous illness', he argues that 'it is one of the obvious social injustices that the standard of civilization should demand from everyone the same conduct of sexual life – conduct which can be followed without any difficulty by some people, because of their organization, but which imposes the heaviest psychical sacrifices on others' (SE IX: 192). It's as if Freud were asking: why should sexual life be governed by a moral code when other areas, for instance eating, are not legislated in this way? It is this dimension of psychoanalysis that allowed it to play such a significant part in the sexual revolutions of the early twentieth century: undermining a repressive system, whether sexual or political, is a radical act.

From its very beginnings, psychoanalysis put forward an idea of the inapplicability of the moral code to the domain of the sexual. It argued that the requirement of abstinence was damaging. It interrogated the knock-on effect of restrictive moral and social codes, including the institution of marriage.

From this perspective, one might say – as Freud does somewhat ironically – that adultery is the cure for neuroses, promiscuity the cure for the constrictions of fidelity. Freud is all but offering immorality as a hygienic recommendation. Psychoanalysis becomes a quasi-medical discourse at war with traditional ethical and social precepts. It is also a discourse that breaks down the distinction between the pathological and the normal, as well as the equation of the pathological and the immoral. For Freud, *all* actions – even exemplary or altruistic ones – can contain pathological or perverse roots. Freud offers an alternative rational account that rejects diagnosis of 'mad' or 'bad' and provides a rational critique of social rules as inherently blind, oppressive and often unfounded.

Because of this, the early history of psychoanalysis is often told as if the psychoanalytic movement is politically revolutionary and aligned with the other radical movements of the period, socialist and Marxist. The interesting alliance that emerged in the 1920s between Marxism and psychoanalysis lasted until the 1970s. Certainly in the view of some of Freud's early followers, such as Otto Gross and Wilhelm Reich, psychoanalysis provides the basis for a vision of a post-revolutionary utopia. For the social theorists of the Frankfurt School from the late 1920s on, Marx and Freud offer complementary revolutionary doctrines. Marxism speaks in the language of classes and workers, but its theory provides no account of individual oppression. Psychoanalysis offers precisely this account of oppression, foregrounding the institutions of family and school, and zeroing in on sexual and gender relations.

The complex interrelations between Marxism and psychoanalysis unfurl in various directions over the course of the century. The Frankfurt School marks a high point of their integration both theoretically and on the streets. When the second edition of Herbert Marcuse's *Eros and Civilization* – a direct hearkening back to Freud's *Civilization and Its Discontents* – came out in the United States in 1966, with a new 'Political

Preface' insisting that the battle against repression and 'the fight for Eros is the *political* fight', it fed directly into the revolutionary fervour and uprisings that marked the late 1960s. Marcuse linked the Freudian idea of the necessary repression of the instincts in the building of civilization to the repressions of advanced industrial capitalism. He argued that the end of sexual and capitalist repression would bring different human relations to being, to nature and instinct. Marcuse's deliberate elision of Marx and Freud's rather different understandings of repression was strategic and came to characterize the historical moment and its language of liberation.

Earlier Frankfurt School theorists had also enlisted psychoanalytic ideas in their attempts to explain Nazism and mid-century fascism. For them, psychoanalysis provided a theory of the means by which irrationality comes to dominate society. The writings of Freud and his followers offered ways of thinking about race and authority, identification with oppressors, and the idealizing of or narcissistic identification with leaders. Quasi-psychoanalytic terms were used to explain the misogynistic and sadistic dimensions of Nazism. Fascist behaviour was understood in terms of perverse character types. Reich's 'little man' hypothesis and his account of a relay between family and class relations leads to the idea of a Freudo-Marxist project in these years. Meanwhile, psychoanalytic ideas about child development and parenting were important to Theodor W. Adorno, the lead writer in *The Authoritarian Personality* (1950), a radical attempt to probe the origins of fascisms and the minds of its followers. Psychoanalytic ideas were also used beyond the left, as psychoanalysts worked for governments to explain the sources of mid-century ideologies, with secret profiles of Hitler commissioned – as Daniel Pick has described in his *The Pursuit of the Nazi Mind: Hitler, Hess and the Analysts* (2012).

This Freudo-Marxist project of the mid-twentieth century had several afterlives. One important one was filtered through the divergent forms of feminism that debated the relevance of psychoanalysis to accounts of family and class relations in the aftermath of the turbulence of 1968. The conflict here was between socialist feminism and psychoanalytic feminism. Socialist feminism from the 1970s on was sceptical of what it saw as the bourgeois tendencies of psychoanalytic feminism. Meanwhile, feminists of varying persuasions, often highly critical of Freud, took up other registers of psychoanalysis – from Klein, to object relations, to Lacan – both practically and in theoretical terms. Many feminist accounts of relations to 'others' and to mothers, to the family and the male gaze, to hysteria as a rebellion against patriarchy and the ambiguities of power relations, drew from a revolutionary psychoanalysis and from efforts at Freudo-Marxist synthesis.

The conservatism of psychoanalysis

Psychoanalysis has also been seen as an inherently reactionary doctrine, one that supports the status quo. Critics argued that its focus on inner conflict offered up an alibi to those who didn't want to change external material conditions. Psychoanalysis made it all too easy to say that oppression was all in the mind and not in actually existing social structures of class exploitation and racial domination and so habituated people into passive acceptance. Rather than interrogating the reality of sexual oppression and violence against women, psychoanalysis allowed a focus on the world of internal fantasy. For many commentators, psychoanalysis's most reactionary form was to be found in the 'ego psychology' that emerged in the US in the 1950s. Here, the basic criterion for health was 'adaptation' to society. The question arises, and it was certainly one Freud had in mind: is there anything 'healthy' in adapting to a pathological society? Critics saw this capitulation to the reactionary ideological constraints of

American society in the post-war period as an indictment of psychoanalysis.

Another reason for characterizing psychoanalysis as reactionary can be found in its equal criticism of all thought systems. Psychoanalysis can be as critical of utopianism, revolutionary or not, as it is of conservative thought. Freud's writings allow all thought systems – religious to revolutionary – to be taken back to an infantile source. This allows rebellious protesting ideologies to be dismissed as infantile and reduced to a relation to a parent – say, the love or murderous rage felt for the authoritarian father. Discussing 1950s American Freudianism, Philip Rieff wrote: 'The new freedom leads to a certain calculated conformity; psychoanalysis finds no more legitimate reasons for being rebellious than for being obedient.'[4] A different way of describing the reactionary political aspect of psychoanalysis can be located in its moral position, which is a fundamentally stoical one (Freud, certainly in his later life, was marked out by his stoicism, his uncomplaining endurance of illness). Critics of psychoanalysis's reactionary or conservative ethos might rightly argue that a stoical ethic, an ethic of endurance, can only be arrived at from a position of privilege, and so provides no ethics at all. Psychoanalysis thus appears as unpleasantly compatible with the interests of a sector of society who are not going to complain about the distribution of privileges: it is the moral stance of the privileged.

Yet in this sense, psychoanalysis shares a perspective with any naturalistic account of human society. Darwinism notoriously struggled to find a naturalistic basis for morality. This gave rise to interminable discussions of the roots of altruism, as though naturalistic proof of altruism would let Darwinism off the hook. But there is no getting off the hook since there is no naturalistic foundation for morality. Even altruism isn't altruism; one may be conditioned to act in the interests of other group members. But though this might look like moral behaviour, it does not amount to morality: it is the self-interested behaviour of the group. Freud accepted this view. He was a hardline naturalist, believing that science offered a replacement for both religion and morality. In this sense, he was just as radical a thinker as Darwin. Indeed, Freud was perhaps even more explicit than Darwin, who strategically cut corners to avoid making enemies. Freud went so far as to liken ethics to table manners or the highway code – an arbitrary system for regulating human interaction: there is no fundamental reason to drive on the left: the reason is practical – to avoid collision. The same might be said, Freud believed, for morality.

Psychoanalysis has also come under heavy criticism as a conservative force in the debate over the place of women in society. At various points in the twentieth century, psychoanalysis was mobilized to bolster regressive 'family values', to argue that women should remain in the sphere of domesticity and child-rearing. This was the case in 1950s America. In Britain too, analysts like D. W. Winnicott – with his ideas of the mother–baby couple, 'good enough' mothering and maternal designations for analysis such as 'holding' – and John Bowlby, have been criticized for disempowering women who had worked through the war years and for promoting an ideology that returned them to the home. Though Freud himself changed his ideas about femininity (and the girl's trajectory through the Oedipal phase) throughout his life, he was certainly Victorian in his first attitudes, believing that many of women's early forms of neurosis might resolve themselves with the birth of a child. Feminist psychoanalysts attacked him for describing women as failed males, lacking the stringent male superego that would mould strong morality.[5] Women also lacked the sublimated homosocial bonding which Freud believed was characteristic of institutional character – from the military to the church.

The argument that women lacked a strong superego (and therefore standard rule-bound morality) has been taken in various different directions. It was harnessed by Carol Gilligan in the 1980s for a feminist purpose. Her book *In a Different Voice* (1982) argued that, far from a deficit, woman's psychology created an alternative form of morality more suited to compassionate relations between human beings. This, Gilligan contended, is what women do best; they don't argue about absolute rules, but rather bring to bear sensitive, situation-based codes of compassion and justice. This is what Gilligan meant by the 'different voice' of her title. Contrast this post-Freudian feminism with earlier nineteenth-century feminisms, which regarded women as asexual guardians of the moral order: both in fact end up characterizing women as superior moral beings by deploying the more reactionary dimensions of psychoanalytic accounts of women as distinctive, but they arrive there by very different routes.

Psychoanalysis as liberalism

As we have seen, psychoanalysis can be criticized from both left and right and has also been used to bolster attacks *on* the politics of both left and right. There is a strong sense in which psychoanalysis can be characterized as a liberalism, in both theory and practice. Freud's view of the human subject undermines the possibility of authority, of certainties, of fully rational choice. The rational subject is at the mercy both of an unpredictable unconscious and the occluded memories of the powers that shaped him from earliest life. Freud thus undermines the possibility of being an uncontested authority in one's own house. This is a view that can be at once both revolutionary but also foundational to a kind of liberalism of uncertainty. When conceived not just as a set of ideas but as a practice, psychoanalysis also looks like a liberalism in another way. It depends for its conditions of possibility and survival on its alliance with social liberalism.

This was a point made by the Viennese scholar and analyst Kurt Eissler, close to Freud and to his daughter Anna, who moved to the US in 1938. In 1974, he noted:

> The psychoanalytic situation is a model situation, which is essentially historically not subject to variation. As far as I can see, only two societal factors must be fulfilled: first, the confidentiality of the analysand's communications must be guaranteed beyond any doubt – that is to say, no government power must be given the right to force an analyst to testify in matters concerning his patient (this possibly makes the practice of psychoanalysis under a dictatorship impossible). Secondly, a certain level of self-observational ability must have been reached within society.[6]

For Eissler, confidentiality is the underlying condition for the very possibility of psychoanalysis. Psychoanalytic practice requires absolute freedom of speech in a context of absolute privacy.

In this sense, psychoanalysis might be thought of as a pure form of liberalism, with the protection of the individual's right to freedom at its core. Psychoanalysis, Eissler argued, is not possible in a totalitarian state. Though attempts were made to form an association purified of Jews in 1930s and 1940s Germany, psychoanalysis was not possible under Naziism. In the Soviet Union, psychoanalysis, which was active during the early revolutionary years, closed down with Stalinization. Similar points could be made about later periods and states: in Brazil and Argentina, during various post-war epochs of dictatorship, the existence of psychoanalysis was certainly challenged, and the behaviour of various psychoanalytic societies and groupings – their collaborative 'leakages' to authority – is only now being opened up for scrutiny.

This argument about the impossibility of psychoanalysis in a totalitarian state can be used to diagnose repressive political changes today; the slippage towards repressive authority can be recognized in the diminishing of the possibility of free unpoliced speech in analysis. The mandatory reporting laws that began to be instituted in the US on therapists in the late 1970s are, in this sense, an index of state repression. These were first implemented after the 1969 Tarasoff murder case: a foreign graduate student at Berkeley at the University of California told his therapist about, and then enacted, the fantasized murder of Tatiana Tarasoff, a woman he thought had encouraged and then rejected him. A doctor's first duty in this new 'reporting' scenario shifted from the patient to the prevention of 'public peril'; the protective privilege of confidentiality was under attack.[7]

Since the 1980s, the permanent crisis surrounding the sexual abuse of children, seen as the fundamental sexual transgression of modern times, has also led to a number of mandatory reporting laws elsewhere.[8] These kinds of policies make psychoanalysis a challenged profession – even an impossible one – in the sense that its fundamental obligation to confidentiality is continually under threat. The psychoanalytic frame of confidentiality is after all meant to protect the borders of fantasy and reality around sexual perversion. In this liberal vision of psychoanalysis, it is perhaps only possible in a free society of a rather delicate kind. This raises a question that we don't have time to consider here, which is whether psychoanalysis is possible outside a Western liberal culture – whether the idea of the subject it posits makes sense outside individualist cultures, and whether its practical programme can be instantiated in non-liberal societies. This, in turn, raises a further question – about whether psychoanalysis is therefore a 'colonialist' bearer of this Western liberal culture, or whether it can be decolonized (a question that concerned Frantz Fanon, as well as the many psychoanalytic theorists who deal with problems of race, colonialism and Blackness).

Psychoanalysis and culture

This marks a good point from which to segue into a look at the more general cultural positioning and influences of Freud and psychoanalysis through the course of the twentieth century and into our own. There are many ways to tell the story of the history of psychoanalysis, some of which we discussed in the last lecture. As a history of ideas, the story of the Freudian century is vast, even if told through the reception of Freud's work alone. If you factor into this picture the influences of other important figures in the field, it's truly enormous: from Winnicott to Lacan and the whole edifice of French psychoanalysis, to literary and cultural theory, from Walter Benjamin and Jacques Derrida to Judith Butler – each of these has a complex reception history. So instead of attempting such a history, let me take here a series of snapshots of psychoanalytic cultural influence in twentieth-century literature, ethics and film. Together, these trace different aspects of psychoanalytic culture – from psychoanalysis's novel conception of an embodied, sexual subject, to the shaping importance of childhood and the overlaps between writing the self and (self-)analysis. But they also all return us again to the central preoccupation of these lectures: to the question of what kind of a thing – and what kind of a practice, system and movement – psychoanalysis is, what it produces and what it does.

The poet W. H. Auden, born in the UK, but because of his years in the US now often thought of as an Anglo-American, was alive to both Marx and Freud when he started writing in the 1920s. Here's a characteristically witty passage – in this case from his brilliant essay 'Psychology and art':[9]

To the man in the street, 'Freudian' literature would embody the following beliefs:

(1) Sexual pleasure is the only real satisfaction. All other activities are an inadequate and remote substitute.
(2) All reasoning is rationalisation.
(3) All men are equal before instinct. It is my parents' fault in the way they brought me up if I am not a Napoleon or a Shakespeare.
(4) The good life is to do as you like.
(5) The cure for all ills is (a) indiscriminate sexual intercourse; (b) autobiography.

Auden's idea that writing about oneself has the same causal force as having sex when it comes to therapy is very well observed, since it's not always clear if speech/language or sex holds greater importance for psychoanalysis. Writing in the 1930s, Auden notes how psychoanalysis has filtered into – even become a symptom of – the general literary, educated, middle brow culture of his period.

Let's look at some other examples of the symptomatic percolation of psychoanalysis in these years.[10] In *A Sketch of the Past*, written in 1940, the year before she died and not long after she had met Freud in London, Virginia Woolf looked back and noted, as if to give further credence to Auden's cue on the uses of autobiography:

> Until I was in my forties . . . the presence of my mother obsessed me. I could hear her voice, see her, imagine what she would do or say as I went about my day's doings . . . in spite of the fact that she died when I was thirteen . . . Then one day walking around Tavistock Square I made up, as I sometimes make up my books, *To the Lighthouse*, in a great, apparently involuntary rush. One thing burst into another. Blowing bubbles out of a pipe gives the feeling of the rapid crowd of ideas and scenes which blew out of my mind, so that my lips seemed syllabling of their own accord as I walked. Why then? I have no notion. But I wrote my book very quickly; and when it was written, I ceased

to be obsessed with my mother. I no longer heard her voice; I do not see her. I suppose I did for myself what psycho-analysts do for their patients. I expressed some very long felt and deeply felt emotion. And in expressing it I explained it and then laid it to rest.[11]

An example of a rather differently inflected cultural percolation of psychoanalysis is evident in George Orwell's important novel *1984* (1949). Its hero Winston Smith's rebellion against the totalitarian society he inhabits is unleashed by Julia, the woman with whom he falls in love. Their prohibited sexual relationship becomes both the generative force and the sign of their dawning individuality and growing defiance against Big Brother. The breaking of their relationship – through Winston's torture – is used in the novel to reveal the power of the state to overcome all resistance and rebellion against it. During one of Winston and Julia's clandestine meetings, she proposes a theory that positions the deformations of sexuality at the very centre of this repressive society:

With Julia, everything came back to her own sexuality. As soon as this was touched upon in any way, she was capable of great acuteness. Unlike Winston, she had grasped the inner meaning of the Party's sexual puritanism. It was not merely that the sex instinct created a world of its own that was outside the Party's control and which therefore had to be destroyed if possible. What was more important was that sexual privation induced hysteria, which was desirable because it could be transformed into war-fever and leader-worship. The way she put it was:

'When you make love, you're using up energy; and afterwards you feel happy and don't give a damn for anything. They can't bear you to feel like that. They want you to be bursting with energy all the time. All this marching up

and down and cheering and waving flags is simply sex
gone sour. If you're happy inside yourself, why should
you get excited about Big Brother and the Three-Year
Plans and the Two Minutes Hate and all the rest of their
bloody rot?'

That was very true, he thought. There was a direct intimate
connexion between chastity and political orthodoxy. For how
could the fear, the hatred, and the lunatic credulity which the
Party needed in its members be kept at the right pitch, except
by bottling up some powerful instinct and using it as a driving
force? The sex impulse was dangerous to the Party, and the
Party has turned it to account.[12]

According to Julia, it is the repression of sexuality and the
re-channelling of sexual energy that allow the society of 1984
to persist.

The Freudian emphasis here has an intriguing echo at the
end of the novel in room 101, the site where souls are broken.
Winston Smith is threatened with torture: ravenous rats will
devour his eyes. Confronted with this, he says: 'Do it to Julia',
a complete betrayal. The image of the rats may well have been
inspired, albeit unconsciously, by Freud's case history of the
Rat Man, who was tormented by a story of a Chinese torture
using rats burrowing up the anus. Orwell modifies this to the
eyes, making it slightly more proper. Freudianism percolated
through the novel in different ways. Of course, Orwell would
not have considered himself a Freudian at all. He was sceptical
of Freud's theories and the whole practice of psychoanalysis,
yet it is hardly a great leap from Julia's ideas about repression
to Freud's, which seem to have a way of seeping into even a
resistant Orwell's vision of a totalitarian society and the make-
up of individuals within it.

If Orwell didn't necessarily have Freud's Rat Man in mind,
the poet Roy Fuller, whose first work, *Poems*, came out in

1939, exemplifies how Freud's case histories, including the Rat Man, invade the literary imagination and shaped views of childhood.[13] Freud's case histories, though they read like approximations of fiction, are here given the status of the real.

> Not the real people in my life
> But the Rat Man and Little Hans
> And other text book creatures have
> Supplied me with romance.
>
> In these the repetitive motives of
> The family were convolved,
> Though I myself from duty, hate
> And affection seem absolved.
> . . .
> To see your father as a horse,
> The penis as what devours,
> Is art without the stigma of
> The artist's ivory towers.
>
> And though the hand that wrote them down,
> To understand and heal,
> Was not entirely guiltless, these
> Strange histories are real.
>
> Yes, real the pince-nez blinkers, real
> The rats that gnaw the heart –
> Emblems of human longing to
> Approximate to art.[14]

The great Vladimir Nabokov thought of Freud as the 'Viennese witch doctor', and considered the therapist as just a space away from the rapist: the therapist is a simpleton and a stereotyper, ready to reduce the (artistic) soul to the merest case. Nabokov understands the work of analysis as an attack on the artist and on individuality, reducing all the threads of the personality to a simplified nonsensical science. And yet psychoanalysis percolates through his work. In perhaps his greatest, certainly his most popular, novel *Lolita*, the frame of the book presents his (anti)hero, Humbert Humbert, as a case, one intent on outwitting the psychiatrist whose case of 'perversion' his story is:

> I discovered there was an endless source of robust enjoyment in trifling with psychiatrists, leading them on; never letting them see that you know all the tricks of the trade; inventing for them elaborate dreams, pure classics in style (which make them, the dream-extortionists, dream and wake up shrieking); teasing them with fake primal scenes; and never allowing them the slightest glimpse of one's real sexual predicament.[15]

I'll just pick out one more, somewhat later, literary manifestation of Freud's ideas, not one of the many novels of Philip Roth, but D. M. Thomas's controversial novel *The White Hotel* (1981), which includes a brilliant pastiche of Freud's letter-writing style. The book is divided into a number of sections, one of which contains fictional letters from Freud about a patient to a colleague. A later section follows a woman in analysis with Freud in 1907; her therapeutic conversations foreground her sadomasochistic fantasies. In the final section, this woman is killed by the Nazis in 1941 at Babi Yar, an infamous site of Holocaust destruction and murder. The sexual fantasies in Thomas's heroine's analysis are thus given 'reality' in the way she is murdered. By aligning the world of a hysteric's sexualized fantasy with the terrible events of the Shoah, by allowing the first to prefigure the second, Thomas stood accused of somehow making Freud complicit with the horrors of the twentieth century. History in the novel's vision becomes a factory for turning fantasies into reality; Freudianism and Nazism emerge as of a piece. Written in the early 1980s, a time when consciousness of the Holocaust was finding a new place in culture, Thomas's postmodern exploration of the hysteric (who shares some features of Sabina Spielrein, the Russian analyst who was Jung's and then Freud's patient, and more than likely also the lover in one way or another of the first) as an avatar of the Holocaust victim rang ominous bells for some critics. Nonetheless, the novel is a sign of the continuing percolations of psychoanalysis in the culture at large.

Psychoanalysis also played a role in a medium that was born around the same time, a medium that has not a little to do with the language of dream: the cinema. The influence of psychoanalytic lore can be clearly seen in 1920s German expressionism: *The Cabinet of Dr Caligari* and the Dr Mabuse films feature destructive hypnotist/somnambulist figures and focus in on the power of the hypnotist-psychiatrist, but also on his closeness to the mad and to the malevolent authority structures that shape a society driven into war.

From the 1930s on in Hollywood, you can see the use of a plot device, borrowed from late nineteenth-century theatre, but here often adapted for the use of a psychoanalytic figure: an outsider – who knows things that other characters don't – transforms the narrative. This outsider – either omnipotent or impotent – serves as a depository of secret knowledge (consider the role of the butler in murder mysteries). Until the end of the twentieth century, this role was often filled by a psychoanalyst – 'the person supposed to know'. The analyst is not directly involved in the basic plotline, but they hear stories and know secrets, thereby standing in for crucial figures of knowledge and power. There are lots of B movie thrillers of this kind. A good later example comes in the famous 1970s film, *The Stepford Wives*. In a perfect and perfectly spooky, suburban town outside New York, women are transformed into robots if they step out of or question their subservient lot as wives and mothers and manifest signs of discontent. In what is an entertaining indictment of American suburban

culture and the place of women – at a time also evoked by the
more recent television series *Mad Men* – the only person who
seems to be in a position to save the heroine from the robotic
fate awaiting her is her psychoanalyst. He tells her to pick up
her children and flee, drive away from this misogynist version
of America. His is the only sympathetic external voice of the
film. But it turns out that she can't, she must succumb to the
power of American ideology. And so, the analyst in America,
in Hollywood, is powerless – a liminal outsider/insider figure.

Among the most famous films to feature psychoanalysis, complete with a stunning dream sequence designed by
Salvador Dali – who had himself visited Freud in London to
pay homage – is Alfred Hitchcock's *Spellbound*. Here there
are four analyst/psychiatrists – three of them working for an
inpatient clinic: Ingrid Bergman, is both the detective and the
patient's love interest; Gregory Peck, the newly arrived head of
the clinic, is also a patient, suffering from amnesia, a suspected
killer and a victim; the third, standing in for greater continental knowledge, is a heavily accented fatherly doctor who
was Ingrid Bergman's analyst/teacher; and finally, Hitchcock
gives us the suave analyst-murderer. The only way for the hero
Gregory Peck to save himself from indictment as the suspected
murderer is to recover his own infantile repressed memories,
which he does with the help of Bergman, the analyst who has
fallen in love with him. The entire plot revolves around the
therapeutic drama seen here as a brilliant piece of detective
work. Freud, the sleuth who knows how to pick the locks of
the psyche, and Sherlock Holmes come together as they do
in many popular culture versions of analysis and sometimes
in Freud's own evocations. Here, there is an explicit analogy
between the thriller detective genre and psychoanalysis.

Behind the various Hollywood representations of psycho-
analysis is the plain fact that many people in the film industry in
Los Angeles – from directors to stars, most famously Marilyn
Monroe – were going through their own analyses behind the

scenes. All those screenplays were being fantasized on the couches of what was called 'couch canyon', North Bedford Drive in LA, though one hopes the casting couch didn't share its more nefarious side with the analytic couch. By the mid-seventies, however, both Hollywood and American culture turn against psychoanalysis. This is in large part because of the rise of psycho-pharmacological drugs and a turn to a more medico-biological understanding of mental health; and in part because of the more general anti-psychiatry movement which, in the first instance, targets the large, notoriously cruel and negligent psychiatric institutions, the 'bins' in which humans are stored, often for life. *One Flew Over the Cuckoo's Nest* (1975), based on Ken Kesey's novel and starring Jack Nicholson, is the anti-psychiatric tour de force of this period. By the time we get to *Silence of the Lambs* in 1991, there is an ultimate Hollywood joke at psychoanalysis's expense: the analyst in this new version now not only kills but eats his victims. But the film and TV industry couldn't stay away from psychoanalysis for long: therapy comes back with the new century in a round of television series, most famously *The Sopranos* and *In Treatment*.

Psychoanalysis and modernity

Let's move now into the portrayals of the development of psychoanalytic culture that we find in the human sciences. I want to take four examples of thinkers who explore the relationship of psychoanalysis to modern conceptions of the self. If we want to understand America's mid-twentieth-century love affair with psychoanalysis, the most interesting voice remains that of Phillip Rieff in his *Freud: The Mind of the Moralist* (1959). In fact, we can't be sure Rieff wrote it himself: he was married to Susan Sontag at the time, who revealed before she died that Rieff had writer's block in the first years of their married life. He had done a great deal of research and thinking, but it was left to Sontag to do much of the actual writing for him. Whoever the author, the book remains important – and brilliantly written. In it, Rieff singles out the role of psychoanalysis in presenting America with a new moral code. He calls this the ethic of honesty. This is, in fact, an anti-ethic, as you'd expect from something coming out of psychoanalysis. For Rieff, psychoanalysis – bound up with the prestige of science – also presents a new iteration of an old scientific ideal: the Enlightenment ideal of founding a social order on a scientific basis. Psychoanalysis provides a version of this. Rieff argues that it 'restored to science its ethical verve' by promoting the idea that *everyone* has an interesting life and the potential for exploring it. Describing mid-century America's fascination with psychoanalysis, Rieff writes that, 'bereft of religion and betrayed by the spurious objectivity of so many sciences, the modern mind has found nothing so convincing as a science that is at the same time a casuistry of the intimate and everyday life'. Freud's science has a double aspect: he provides an indictment of modern society in the form of a science which denies its own status as a moral science: 'The illness it proposes to treat scientifically is precisely our inherited morality, and therefore it is . . . both a natural and a moral science.'

Rieff delineates the structures of civilizations to show the importance of psychoanalysis to modernity, presenting the movement of civilizations and leading types to give the age of therapy its name. The Romans have the philosopher, the Christians have the pastor; the psychoanalyst is the equivalent figure for the dispirited modern:

> As the aristocratic Roman summoned his philosopher when he was ill, as the Christian went to his pastor, so the dispirited modern visits his analyst. But the psychoanalyst does not compete with older therapists; his is not a therapy of belief but one which instructs how to live without belief . . . Religion can no longer save the individual from forming his private neurosis,

for he has become his own religion: taking care of himself is his ritual now, and health is the ultimate dogma.[16]

For Rieff, one of the distinctive features of psychoanalysis is that it is a moral system that does not require belief. In other words, therapy, though it performs a similar function to religion, fundamentally subverts it. Here Rieff is advancing the notion of the era of 'psychological man', whom he contrasts with familiar historical types: the 'political man', the ideal figure of public action and virtue of the Ancient world; the 'religious man' of the Judeo-Christian tradition, who denies the importance of this world in favour of the salvation of one's soul for the other world; and 'economic man', a product of the Enlightenment, of Adam Smith's 'rational-economic man' and the Humean 'calculating man', who segues into contemporary neoliberalism and rational choice theories of the late twentieth century. For Rieff, the Freudian 'psychological man' represents a shift towards an ideal of insight that can lead to self-mastery, an ability to live beyond the reason of 'political man', the conscience of Christianity, and the optimistic rationalized utopias of 'Enlightenment man'. He views the era of psychological man with ambivalence:

> So far as it clears away the residual identities that make people ill, Freudianism does not thereby destroy its own capacity to forge a new identity, for it helps bring to life what is essentially a counter-identity. Once the patient is free of the last authority, the therapist, he has achieved the only possible and real freedom; he is himself alone – however diminished that self may be. Thus are disengagements masked as liberations, modest encouragements to life for its own sake proposed as an ideal of freedom.[17]

A similar account of 'characters' is presented in the Scottish-American philosopher Alasdair MacIntyre's influential book,

After Virtue (1981). This introduces the notion of 'character' to portray the moral ambience or ethos of an era. Akin to stock characters in plays, these 'characters are the masks worn by moral philosophers'. Characters, for MacIntyre, are:

> a very special type of social role which places a certain kind of moral constraint on the personality of those who inhabit them in a way which many other social roles do not . . . In the case of a character, role and personality fuse in a more specific way than in general . . . One of the key differences between cultures is in the extent to which roles are characters; but what is specific to each culture is in large and central part what is specific to its stock of characters. So the culture of Victorian England was partially defined by the characters of the Public School Headmaster, the Explorer and the Engineer; and that of Wilhelmine Germany was similarly defined by such characters as those of the Prussian Officer, the Professor and the Social Democrat.[18]

Moral life is dominated by such characters, who give the moral tone to a society. 'A *character* is an object of regard by the members of the culture generally or by some significant segment of them', MacIntyre continues. 'The *character* morally legitimates a mode of social existence.'[19]

MacIntyre's list of twentieth-century characters includes the rich aesthete (the hedonist), the manager and the therapist. What's important to MacIntyre – who is himself a moralist committed to an old-fashioned morality – is that these characters avoid true moral discourse by a recourse to a non-moral rationality: 'The manager represents in his character the obliteration of the distinction between manipulative and non-manipulative social relations; the therapist represents the same obliteration in the sphere of personal life . . . Neither manager nor therapist, in their roles as manager and therapist, do or are able to engage in moral debate.'[20] This peculiar twentieth-century phenomenon marks for MacIntyre an undercutting of moral discourse by these new characters who embody an anti-moral moral ethos. This undercutting is what the Freudian therapist represents.

The anthropologist and political analyst Ernest Gellner, in his exhilarating and witty landmark book *The Psychoanalytic Movement* (1985), sees psychoanalysis as a stand-in for religion in an irreligious world, 'a secular procedure with authority of science that offers hope to those in trouble with their fellow humans'. The twentieth century for him is marked not by human confrontations with nature as in earlier periods, but with other human beings – though perhaps we are returning to a version of the first now. Cleverly subtitled *The Cunning of Unreason*, to depict the logic of the unconscious, Gellner's study accounts for the ways in which psychoanalysis became so dominant in the twentieth century at the moment in the history of civilization when these confrontations between human beings become so central to culture. As an anthropologist, Gellner seeks to understand why psychoanalysis is so successful. He points to Freudian psychoanalysis's ways of insulating itself from criticism and establishing a set of social bonds that become similar to a religion and that emerge through the transference relationship. He also finds that it responds to a set of cultural needs, as specified by Nietzsche, for an accurate account of what human life is like.

Gellner argues that, despite this, psychoanalysis is the one quasi-scientific discourse that responded to the new historical and cultural pressure of the end of the nineteenth century. He calls this the 'Nietzsche minimum': psychoanalysis meets the conditions of the minimum set of requirements first postulated by Nietzsche as necessary to an adequate account of what human beings were like in the new cultural circumstances of modernity. By contrast, other conceptions fail to capture reality, or have now become irrelevant. Gellner draws a comparison with eighteenth-century psychology – with atomistic conceptions of human beings, including Hume's 'bundle' conception of the self, according to which the self is 'nothing but a bundle or collection of different perceptions, which succeed each other with an inconceivable rapidity, and are in a perpetual flux and movement'. Today, this conception is inappropriate, and out of line with any description of the reality of our actual lives. In the twentieth century, an adequate account of the self must account for our awareness of our instinctuality, of the bodilyness of our being; our situationality, and the here-and-nowness of our own experiences; our experience of reality as other people – particularly our intimate relations of power and submission. It must account for the fact that we describe our lives and experience in a fundamentally gestalt and traumatic mode; we have an immediate traumatic perception of situations (contrast this with a Humean or utilitarian vision, which adds up the pros and cons of courses of action over a period of time – the accountancy procedures of empiricist philosophy). It must also capture the fact that much of our selves is not accessible to consciousness or deliberate control, and the fact that infancy is a crucial crystallization period, among other criteria.

For Gellner, the success of psychoanalysis is found in its capacity to meet these conditions. But, on the other hand, Gellner wants to say that psychoanalysis is as bad, if not worse, than the old religious institutions, since it protects itself by building in an arcane set of social relations. People are never allowed out of psychoanalysis once they're in because of transference relations. This gives it the character of a religious cult dressed up as a science, but one that simultaneously responds to a real set of needs in Western society at the beginning of the twentieth century.

Gellner's picture of psychoanalysis and its percolation in culture is complex. I want to close today with the equally complex account of psychoanalysis provided by Michel Foucault. There is no time to address these complexities adequately here and I will be giving another set of lectures next term on Foucault and the history of the human sciences. I just want to say now that, in his earlier work, Foucault thinks of psychoanalysis as one of the most radical intellectual machines for establishing a human science. In *The Order of Things*, he posited that the unconscious provided the underbelly of the modern episteme, its condition of possibility. But by the time he writes the first volume of *History of Sexuality*, and is thinking about sexuality, he is primarily critical of psychoanalysis.

Foucault argues that political discourse, as it had emerged from psychoanalysis and other discourses of sexuality, had emphasized repression. It had set up a model of 'truth', according to which the 'truth' of the individual is opposed to 'power'. Repression had been cast as the main kind of power that operates in the field of sex; the point was to seek a critical analysis of – and liberation from – that repression. But Foucault asks a sceptical question about this 'repressive hypothesis':

> The question I would like to pose is not, Why are we repressed? but rather, Why do we say, with so much passion and so much resentment against our most recent past, against our present, and against ourselves, that we are repressed? By what spiral did we come to affirm that sex is negated? . . . What paths have brought us to the point where we are 'at fault' with respect to our own sex?[21]

Psychoanalysis, with its discourse of sex, was among his key targets.[22]

What Foucault saw, as a historian, much more straightforwardly, was that psychoanalysis was entirely complicit with the incitation to sexual speech: it falls in with the new modes of discipline introduced at the beginning of the nineteenth century that came to fruition in the twentieth. In his account of the development of the human sciences – and particularly of the general economy of discourses about sex that had developed since the seventeenth century and more recently, with the rise of criminology, psychiatry and sexology – psychoanalysis becomes the embodiment of new oppressive forms of incitement to sexual discourse, which Foucault, writing in the context of the new identity politics of the 1970s, wanted to call attention to.

Providing a genealogy of interpretation, Foucault shows that psychoanalysis lies within the tradition of confession. This is true in two senses. First, it is the inheritor of Christian interpretative, hermeneutic, and confessional practices – those practices directed at the hidden state of the soul. Here, the Christian and Freudian efforts to reveal secrets – secrets that are the sources of hidden truth – are analogous. To recover practices that were not dominated by the Christian and Freudian drive for hidden truth, Foucault returned to the Greeks and Romans to provide the basis for a different ethics of sexuality and the body. Second, Foucault also posits an historical thesis about psychoanalysis and confession: that confession, detached from the sixteenth-century sacrament of penitence, migrated through pedagogy to the family, to medicine and psychiatry. These disciplines both produce knowledge of individuals and new kinds of individuals. The novel forms of individualizing power – like the powers of medicalization and examination – produce new kinds of people; in his most famous example, the homosexual is produced as a new character, a new personage, with a new past. In the regime of power-knowledge that Foucault sees as being born with the disciplines and the sexual sciences, the examination, like the case, consigns and confines the individual; individuals do not speak their truths to power (or speak their truths about sex). Rather, they are produced and defined by power operating in a range of modes – the control of infantile sexuality, the division and policing of normal and abnormal, the power of medicalization.

On this view, the way to become or to be acknowledged as an individual in the twentieth century is to become a case. To close where we began, psychoanalysis, here, is indeed a science. Yet it is not a scientific movement, but a discipline, with power operating through it. With its case histories, its search for latent truth, psychoanalysis is, for Foucault, part of the human sciences and the *scientia sexualis* that construct and produce individuals. Writing in the 1970s, this characterization of psychoanalysis would become a staple feature of anti-psychiatry culture. In many ways, Foucault, with his emphasis on discipline and the body and his revaluation of science and power, presages the decline of psychoanalysis. For him, truth is no longer found in the relationship to the other or to the transcendent. Whether psychoanalysis – the impossible profession – can, or will, survive that inversion of truth remains to be seen.

We should, however. give the last word to Freud, who was clear that the impossible profession that was psychoanalysis depended on the absolute privilege of psychoanalytic speech:

> It is very remarkable how the whole [analytic] task becomes impossible if a reservation is allowed at any single place . . . I once treated a high official who was bound by his oath of office not to communicate certain things because they were state secrets, and the analysis came to grief as a consequence of this restriction. Psycho-analytic treatment must have no regard for any consideration, because the neurosis and its resistances are themselves without any such regard. (SE XII: 135–136n)

Photograph of Sigmund Freud and dogs taken at Hohe Warte 46, Vienna, Summer 1933. © Freud Museum London

Editor's Postscript

John Forrester ended these lectures abruptly, with a mournful ambivalence about psychoanalysis. But he was ambivalent in the psychoanalytic sense of the word; he did not always take Foucault's side.

He ended another of his introductions to psychoanalysis in a different way, in a more optimistic mood. The closing remarks of his introduction to the Penguin edition of *Interpreting Dreams* provide an alternative set of last words that stressed the promise and possibility at the heart of the psychoanalytic project. Psychoanalysis, he wrote there, can also be seen as 'a model for making an extraordinary life out of the ordinariness of the everyday':

> The heroic and the bestial are both simultaneously deprived of their magic and made the property of all; each of us, through following our dreams as Freud did, can discover the 'excitement of the wholly interesting life'. Freud presages the possibility of a democracy of the inner life to follow on the heels of the democracy of suffrage and education . . . Everyone has an inner life and the right, if not always the chutzpah, to share it.[1]

Notes

Lecture 1: A Whole Climate of Opinion

1 W. H. Auden, 'In memory of Sigmund Freud', © 1940.

2 'Sigmund Freud BBC speech', https://www.freud.org.uk/2020/07/13/freuds-bbc-speech/.

3 https://worldhistoryproject.org/1930/8/28/sigmund-freud-is-awarded-the-goethe-prize.

4 David Goodstein, 'In defense of Robert Andrews Millikan', *Engineering & Science*, 4 (2000). http://calteches.library.caltech.edu/4014/1/Millikan.pdf.

5 E. G. Boring, *A History of Experimental Psychology*, 2nd edn. (New York: Appleton-Century-Crofts, 1950[1929]).

6 Kurt Danziger, *Constructing the Subject: Historical Origins of Psychological Research* (Cambridge: Cambridge University Press, 1990).

7 Paul Ricœur, *Freud and Philosophy*, trans. Denis Savage (New Haven, CT: Yale University Press, 1970), p. 33 and passim.

8 'Psycho-analysis is the name (1) of a procedure for the investigation of mental processes which are almost inaccessible in any other way, (2) of a method (based upon that investigation) for the treatment of neurotic disorders and (3) of a collection of psychological information obtained along those lines, which is

gradually being accumulated into a new scientific discipline' (SE XVIII: 235).

9 *The Complete Letters of Sigmund Freud to Wilhelm Fliess 1887–1904*, trans. and ed. Jeffrey Moussaieff Masson (Cambridge, MA: The Belknap Press, 1985), p. 398.

10 Austin differentiated between *constative* and *performative* utterances. See J. L. Austin, *How to Do Things with Words*, 2nd edn., ed. M. Sbisà and J. O. Urmson (Oxford: Oxford University Press, 1975[1962]).

11 Robert Linder, *The Fifty-Minute Hour* (New York: Rinehart & Company, 1955; repr. London: Free Association Books, 1956).

12 Lindner, *The Fifty-Minute Hour*, 1956, p. 291.

13 Freud to Sándor Ferenczi, 13 October 1910, in *The Correspondence of Sigmund Freud and Sándor Ferenczi, Vol 1: 1908–1914*, ed. Eva Brabant, Ernst Falzeder and Patrizia Giampiere-Deutsch, trans Peter T Hoffer (Cambridge, MA: Harvard University Press, 1993).

14 See Freud, 'Notes upon a case of obsessional neurosis' (1909), SE X: 191; 'The disposition to obsessional neurosis' (1913), SE XII: 320.

15 The Vienna Circle was a group of early twentieth-century philosophers who sought to reconceptualize empiricism by means of their interpretation of then recent advances in the physical and formal sciences. https://plato.stanford.edu/entries/vienna-circle/.

16 The 'naturalists' of the early twentieth century include John Dewey, Ernest Nagel, Sidney Hook and Roy Wood Sellars. These philosophers aimed to ally philosophy more closely with science. They urged that reality is exhausted by nature, containing nothing 'supernatural', and that the scientific method should be used to investigate all areas of reality. https://plato.stanford.edu/entries/naturalism/.

17 Psychoanalytic theories were, Popper argues, 'simply non-testable, irrefutable. There was no conceivable human behaviour which could contradict them'. Popper described Marxism as 'reinforced dogmatism'. See *Conjectures and Refutations: The*

Growth of Scientific Knowledge, (London: Routledge, 1963), pp. 37, 334.

18 Vladimir Nabokov, *Strong Opinions* (London: Penguin Classic, 1973), p. 66.

19 Philip Rieff, *Freud: The Mind of the Moralist* (Chicago: The University of Chicago Press, 1959), p. 304.

Lecture 2: The Historical Foundations of Psychoanalysis

1 Ernest Jones, *The Life and Work of Sigmund Freud. Vol 1: The Young Freud* (London: The Hogarth Press, 1953).

2 See Morton Prince, *Psychotherapy and Multiple Personality* (Cambridge, MA: Harvard University Press, 1975).

Lecture 3: Dreams and Sexuality

1 *The Complete Letters of Sigmund Freud to Wilhelm Fliess 1887–1904*, trans. and ed. Jeffrey Moussaieff Masson (Cambridge, MA: The Belknap Press, 1985), pp. 264–265.

2 John Forrester, 'From neurology to philology: language and the origins of psychoanalysis', PhD thesis (University of Cambridge, 1979), published as *Language and the Origins of Psychoanalysis* (London: Macmillan, 1980).

3 Bertrand Russell, *Marriage and Morals* (London: Allen & Unwin, 1929). https://russell-j.com/beginner/MaM1929-TEXT.HTM.

Lecture 4: Psychoanalysis as a Theory of Culture

1 The introduction to this chapter has been reconstructed from notes by the editor.

2 Philip Rieff, *Freud: The Mind of the Moralist* (Chicago: The University of Chicago Press, 1959).

3 *The Freud/Jung Letters*, ed. William McGuire (London: Hogarth Press, 1974), p. 255.

4 Freud's understanding of the ego ideal and narcissism in the formation of groups has been incisively used in analysing the Trump phenomenon, one that post-dates Forrester's lectures. (Editor's note)

Lecture 5: Psychoanalysis as an International Movement

1 This chapter combines text transcribed from Forrester's lectures and reconstructions of unrecorded lecture notes. The introductory paragraphs were written by the editor.

2 Philip Rieff, *Freud: The Mind of the Moralist* (Chicago: The University of Chicago Press, 1959), p. 304.

3 Jacques Derrida, *La Carte Postale* (Paris: Aubier-Flammarion, 1980), p. 324.

4 Freud to Abraham, 8 October 1907, in *The Complete Correspondence of Sigmund Freud and Karl Abraham 1907–1925*, ed. Ernst Falzeder (London, New York: Karnac Press, 2002), p. 9.

5 Edward Shorter, *A History of Psychiatry* (New York: Wiley, 1998).

6 This paragraph was reconstructed from Forrester's lecture notes by the editor.

7 Pat Barker, *Regeneration* (1991); *The Ghost Road* (1993); *The Eye in the Door* (1995).

8 *The Berlin Years: Writings, 1918–1921*, in *The Collected Papers of Albert Einstein*, vol. 7, ed. Michel Janssen, József Illy, Christoph Lehner and Diana K. Buchwald (Princeton: Princeton University Press, 2002), p. 214. https://einsteinpapers.press.princeton.edu/vol7-doc/262.

9 See John Forrester's 'The psychoanalytic passion of J. D. Bernal in 1920s Cambridge', *British Journal of Psychotherapy*, 24/4 (2010), pp. 397–404, (where he cites from the copious Bernal Papers in the University of Cambridge Library (https://www.hps.cam.ac.uk/files/forrester-psychoanalytic-passion.pdf). Also see his and Laura Cameron's *Freud in Cambridge* (Cambridge: Cambridge University Press, 2016).

10 The remainder of this paragraph was added by the editor.

11 Elisabeth Roudinesco, *Jacques Lacan & Co: A History of Psychoanalysis in France* (London: Free Association Books, 1990), p. 197.

12 The following section on Lacan was reconstructed from Forrester's lectures notes by the editor.

13 See much more on this in Forrester, *Freud in Cambridge*.

14 The following paragraphs on Anna Freud, Winnicott and Klein were reconstructed from Forrester's lecture notes by the editor.

15 https://www.cabinetmagazine.org/issues/20/falzeder.php.

16 This paragraph was reconstructed from Forrester's lecture notes by Katrina Forrester.

Lecture 6: The Significance of Psychoanalysis in the Twentieth Century

1 The introduction to this chapter was reconstructed from Forrester's lecture notes by the editor and Katrina Forrester.

 2 Malcolm Bowie, 'Freud and the European unconscious', in David Lodge and Nigel Wood, eds., *Modern Criticism and Theory: A Reader* (London: Routledge, 2014), p. 628.

 3 Keynes, under the pseudonym Suela in *The Nation Athenaeum* on 29 August 1925. See John Forrester, '"A sort of devil" (Keynes on Freud, 1925): Reflections on a century of Freud-criticism', *Österreichische Zeitschrift für Geschichtswissenschaften*, 14/2 (2003), pp. 70–85. https://doi.org/10.25365/oezg-2003-14-2-4.

 4 Philip Rieff, *Freud: The Mind of the Moralist* (Chicago: The University of Chicago Press, 1959), p. 328.

 5 See Lisa Appignanesi and John Forrester, *Freud's Women*, 3rd edn (London: Phoenix Books, 2010).

 6 Kurt L. Eissler, 'On some theoretical and technical problems regarding the payment of fees for psychoanalytic treatment', *International Journal of Psycho-Analysis* 1 (1974), pp. 73–101: p. 74.

 7 For more on this, see Lisa Appignanesi, *Trials of Passion* (London: Virago Press, 2014).

 8 This paragraph was reconstructed from notes by the editor and Katrina Forrester.

 9 W. H. Auden, 'Psychology and art', in Geoffrey Grigson, ed., *The Arts To-day* (London: John Lane, The Bodley Head, 1935).

10 Some of these examples were added by the editor.

11 Virginia Woolf, *Moments of Being*, ed Jeanne Schulkind (New York: Harcourt Brace, 1976), p. 80. Also in Woolf, *A Writer's Diary* (London: Hogarth Press, 1953), entry dated 28 November 1928.

12 George Orwell, *1984* (London: Penguin Classics, 1949), pp. 87–88.

13 The following two paragraphs contain additions to Forrester's lectures made by the editor.

14 Roy Fuller, 'Freud's case histories,' in *Tiny Tears* (London: Andre Deutsch, 1973).

15 Vladimir Nabokov, *Lolita* (New York: Crest Books Edition, 1959[1955]), p. 34.

16 Rieff, *Freud: The Mind of the Moralist*, p. 334.

17 Rieff, *Freud: The Mind of the Moralist*, p. 333.

18 Alastair MacIntyre, *After Virtue* (London: Bloomsbury, 2007), p. 28.

19 MacIntyre, *After Virtue*, p. 28.

20 MacIntyre, *After Virtue*, p. 10.

21 Michel Foucault, *The History of Sexuality*, vol. 1, trans. Robert Hurley (New York: Random House, 1978), p. 8.

22 The remaining text has been reconstructed from a combination of the recorded lecture and Forrester's lecture notes by Katrina Forrester.

Editor's Postscript

1 John Forrester, 'Introduction,' in *Interpreting Dreams* (London: Penguin Books, 2006), p. xlviii.

Further Reading

Editor's note: The list below is a selection drawn from the annotated reading list that Forrester prepared for students who attended his lectures.

I. Works by Freud

The Standard Edition of the Complete Psychological Works of Sigmund Freud, 24 vols, ed. James Strachey (London: Hogarth Press, 1953–74). Electronic access to the Standard Edition of Freud's writings and nearly all psychoanalytic papers published in English since 1920 at PEP-WEB (http://www.pep-web.org/).

A new set of translations – in a series whose General Editor is Adam Phillips – is available in Penguin Books.

On psychoanalysis

Introductory Lectures on Psychoanalysis (1916–17), SE XV and XVI. The best, but also the longest introductory account. For Freud's additions and emendations, including the ego/super-ego/id schema, see:

New Introductory Lectures (1933), esp. lecture entitled 'The dissection of the psychical personality', SE XXII and PF 2.

Five Lectures on Psycho-Analysis (1910). Short and very clear introduction.
The Question of Lay Analysis (1926), SE XX: 183–258. A very clear
exposition of psychoanalysis for the general reader, which is also a
polemic against doctors and those who conceive of psychoanalysis
as a branch of medicine.

On Dreams

The Interpretation of Dreams (1900), SE IV and V. For a compressed
version, see either the introductory account in *Introductory
Lectures* (Lectures V–XV), or 'On Dreams' (1901), SE V: 633–
686.

On Sexuality

Three Essays on the Theory of Sexuality (1905), SE VII: 130–243.
Introductory Lectures, SE XVI: Lectures XX, XXI and XXVI.
'A special type of object-choice made by men' and 'On the universal
tendency to debasement in the sphere of love', both in SE XI.
'Character and anal-erotism' (1908), SE IX: 169–175.
'The dissolution of the Oedipus complex', SE XIX: 173–179.

Case-histories

Studies on Hysteria (1895). Includes Breuer's first case, Anna O.
'A fragment of the analysis of a case of hysteria' (1905), SE VII. The
Dora case, much commented upon by writers interested in Freud's
views on femininity and transference.
'Analysis of a phobia in a five-year-old boy' (1909), SE X. The case of
'little Hans'; the first child analysis.
'Notes upon a case of obsessional neurosis' (1909), SE X. The case
of the 'Ratman' – includes a blow-by-blow account of the first
sessions of the analysis, and transcripts of some of Freud's unpublished case-notes.
'From the history of an infantile neurosis' (1914/18). The 'Wolfman':
the deepest and most complex of the case-histories, with a lengthy
discussion of the reality of infantile traumas and events.

'The psychogenesis of a case of female homosexuality' (1920), SE XVIII: 147–182. Seemingly slight, but with interesting remarks on homosexuality, deception and unconscious sexuality.

Theoretical and metapsychological

Project for a Scientific Psychology (1895), SE I. Freud's unpublished attempt to give a neurological account of normal and pathological functioning; this is, the 'neuroscientific' Freud. Very complex, but many of his later hypotheses can be found embedded within it, virtually fully developed.

The Interpretation of Dreams (1900), SE V. Chapter VII is Freud's first full-blown theoretical model, with theories of unconscious, preconscious and conscious, and exploration of distinction between primary and secondary processes of mental functioning.

'Formulations on two principles of mental functioning' (1911), SE XII: 218–226. Introduces the distinction between reality and pleasure principles.

'The unconscious' (1915), SE XIV: 166–204. A theoretical essay on Freud's most important concept.

'Repression' (1915), SE XIV. The theoretical essay on the concept Freud called the 'corner-stone of psychoanalysis'.

'Instincts and their vicissitudes' (1915), SE XIV. On the concept of instinct or drive, with complex developments in the theory of love and hate. The introductory remarks outline Freud's philosophy of science.

Beyond the Pleasure Principle (1920), SE XVIII: 7–64. Introduces the compulsion to repeat, and Freud's speculations concerning the love and death drives.

The Ego and the Id (1923), SE XIX: 7–63. Introduces the concept of the id and the super-ego.

Technique and interpretation

'Screen-memories', SE III: 303–322.

'Observations on transference-love', SE XII: 157–171.

'Remembering, repeating and working-through', SE XII: 147–156.

Other papers on technique are also to be found in SE XII, and the last two lectures of the *Introductory Lectures*, SE XVI (Lectures XXVII 'Transference' and XXVIII 'Analytic therapy'), are very useful summaries, raising all sorts of questions about the nature of cure, the relation between psychoanalysis and suggestion, etc.

Other major concepts

'On narcissism: an introduction' (1914), SE XIV: 73–102.

'Mourning and melancholia' (1917), SE XIV: 242–258. Introduction of the notion of object-loss; a very important paper, claimed to be the most cited paper in the history of psychology and psychiatry.

'Negation' (1925), SE XIX: 235–239. A remarkable four pages long, Freud's densest: on why 'no' sometimes means 'yes'.

'Note upon *The Mystic Writing Pad*' (1925), SE XIX: 227–231. A very interesting example of how to set up a scientific model.

Writings on femininity

'The infantile genital organization' (1923), SE XIX: 141–145; PFL 7.

'Some psychical consequences of the anatomical distinction between the sexes' (1925), SE XIX: 248–258; PFL 7.

'Female sexuality' (1931), SE XXI: 225–243; PFL 7.

'Femininity', Lecture XXXIII in *New Introductory Lectures*, SE XXII: 112–135; PFL 2.

Social and political thought

Totem and Taboo (1912–13), SE XIII: 1–161. On the analogy between neurosis and magical and religious rites; includes hypothesis of primal horde and the murder of the primal father.

Group Psychology and the Analysis of the Ego (1921), SE XVIII: 69–143. Includes discussion of intimate relation between hypnosis, love and psychoanalysis; analysis of crowds and leadership, and Freud's fullest discussion of the concept of identification.

The Future of an Illusion (1927), SE XXI: 5–56. Freud's critique of religion as a collective neurosis, nay delusion.

Civilization and Its Discontents (1930), SE XXI: 64–145. Freud's pessimistic view of society, and a lengthy discussion of guilt. The *locus classicus* for the 'Freudian' view of human society.

II. Selected Works on Freud and Psychoanalysis

Works of reference

Anzieu, Didier, *Freud's Self-Analysis*, tr. Peter Graham (London: Hogarth Press & Institute of Psychoanalysis, 1986). A comprehensive account of the interaction of Freud's inner life and the development of his theories in the 1890s.

Ellenberger, H., *The Discovery of the Unconscious* (London: Allen Lane, 1970). Indispensable resource for the history of psychoanalysis, its relationship to hypnotism, psychiatry, occultism, etc.

Gay, Peter, *Freud. A Life for Our Time* (London: Dent, 1988). Generally well judged, with an excellent bibliographical essay on the state of Freud scholarship as of 1988, already vast then.

Jones, Ernest, *The Life and Work of Sigmund Freud* (London: Hogarth Press, 1953–7). Written by the Disciple, but still a classic. It has more well-documented 'facts' than any other biography.

Laplanche, J. and Pontalis, J.-B., *The Language of Psychoanalysis* (London: Hogarth Press, 1973). Definitive and comprehensive dictionary of terms and concepts.

Spurling, Laurence (ed.), *Sigmund Freud. Critical Assessments*, 4 vols (New York: Routledge, 1989). A very useful collection of the vast secondary literature on all aspects of Freud's work.

Major secondary sources

1. Introductory and synoptic overviews

Mannoni, Octave, *Freud. The Theory of the Unconscious* (London: Pantheon, 1971).

Smith, Roger, 'The unconscious: reason and unreason', in *The Fontana History of the Human Sciences* (London: Fontana, 1997), pp. 701–745.

Wollheim, Richard, *Freud* (Fontana Modern Masters, 1970); 2nd edn 1991, with new Introduction.

2. Specialist studies on historical development of psychoanalytic theory

Appignanesi, Lisa and John Forrester, *Freud's Women* (London: Weidenfeld & Nicolson, 1992). New edns: Penguin 2000, Orion 2005.

Borch-Jacobsen, Mikkel, *Remembering Anna O. A Century of Mystification*, trans. Kirby Olson, in collaboration with Xavier Callahan and the author (New York: Routledge, 1996).

Chertok, Leon and Raymond Saussure, 'The birth of psychoanalysis', in Laurence Spurling (ed.), *Sigmund Freud. Critical Assessments*, vol. I (New York: Routledge, 1989), pp. 402–426.

Crews, Frederick C., *Unauthorized Freud: Doubters Confront a Legend* (London: Viking, 1998). A collection of the violent criticisms of Freud as man, scientist, therapist.

Davidson, Arnold I., 'How to do the history of psychoanalysis: A reading of Freud's *Three Essays on the Theory of Sexuality*', *Critical Inquiry* 13/2 (1987), pp. 252–277.

Davidson, Arnold I., *The Emergence of Sexuality: Historical Epistemology and the Formation of Concepts* (Cambridge MA: Harvard University Press, 2001).

Derrida, Jacques, 'Freud and the scene of writing', in *Writing and Difference*, trans. Alan Bass (Chicago, IL: University of Chicago Press, 1978). Derrida's essays on Freud and psychoanalysis are idiosyncratic, difficult and very important; this first one is an analysis of Freud's neurological model.

Derrida, Jacques, 'To speculate – on "Freud"', in *The Post Card. From Socrates to Freud and Beyond*, trans. Alan Bass (Chicago, IL: University of Chicago Press, 1987). A brilliant analysis of Freud's biological speculation, *Beyond the Pleasure Principle*.

Derrida, Jacques, *Resistances of Psychoanalysis*, trans. Peggy Kamuf, Pascale-Anne Brault and Michael Naas (Stanford, CA: Stanford University Press, 1998). Three typically allusive but brilliant essays. Short and sweet.

Derrida, Jacques, 'To do justice to Freud', in Arnold I. Davidson (ed.), *Foucault and His Interlocutors* (Chicago, IL: University of Chicago

Press, 1997). Searching and moving analysis of Foucault's placing of Freud.

Forrester, John, *Language and the Origins of Psychoanalysis* (London: Macmillan, 1980).

Forrester, John, *The Seductions of Psychoanalysis* (Cambridge: Cambridge University Press, 1990).

Forrester, John, *Dispatches from the Freud Wars. Psychoanalysis and its Passions* (Cambridge, MA: Harvard University Press, 1997).

Foucault, Michel, *The History of Sexuality*, vol. I, trans. Robert Hurley (New York: Random House, 1978).

Forrester, John and Laura Cameron, Freud in Cambridge (Cambridge: Cambridge University Press, 2016).

Foucault, Michel, *Power/Knowledge, Selected Interviews and Other Writings 1972–1977*, ed. Colin Gordon (New York: Pantheon, 1980).

Foucault, Michel, 'Sex and truth', in *The Foucault Reader*, ed. Paul Rabinow (London: Peregrine Books, 1986).

Gellner, Ernest, *The Psychoanalytic Movement, or the Cunning of Unreason* (London: Paladin, 1985); 2nd edn, 1993/96. A scathing, marvellously written attack on psychoanalysis, posing important questions about its significance in a secular, therapeutic society.

Ginzburg, Carlo, 'Morelli, Freud and Sherlock Holmes: clues and scientific method', in Umberto Eco and Thomas A. Sebeok (eds), *The Sign of Three: Dupin, Holmes, Pierce* (Bloomington: Indiana University Press, 1984), pp. 81–118.

Hacking, Ian, *Rewriting the Soul: Multiple Personality and the Sciences of Memory* (Princeton: Princeton University Press, 1995). Account of the rise of 'memoro-politics' – the sciences of memory – and the acute politics associated with child abuse, multiple personality disorder of the 1980s and 1990s. Psychoanalysis fills the gap between the late nineteenth century and then.

Laqueur, Thomas, *Making Sex: Body and Gender from the Greeks to Freud* (Cambridge, MA: Harvard University Press, 1990), esp. pp. 227–243.

McGrath, William J., *Freud's Discovery of Psychoanalysis. The Politics of Hysteria* (Ithaca, NY: Cornell University Press, 1986).

Makari, George, *Revolution in Mind: The Creation of Psychoanalysis* (New York: HarperCollins, 2008).

Micale, Mark, *Approaching Hysteria: Disease and Its Interpretations* (Princeton, NJ: Princeton University Press, 1995).

Mitchell, Juliet, *Psychoanalysis and Feminism* (London: Allen Lane, 1974); new edn. 2000.

Phillips, Adam, *Darwin's Worms* (London: Faber & Faber, 1999). Phillips is the author of a series of wise and sensitive readings of Freud and of psychoanalysis for the end of Freud's century and into the twenty-first; see, for example, *On Kissing, Tickling and Being Bored, The Beast in the Nursery* and *Terrors and Experts*, as well as his fine anti-biography, *Becoming Freud: The Making of a Psychoanalyst* (New Haven, CT: Yale University Press, 2014).

Rieff, Philip, *Freud: The Mind of the Moralist*, 3rd edn. (Chicago, IL: The University of Chicago Press, 1975 [1959]). The most penetrating and culturally sophisticated understanding and critique of Freud's vision and of psychoanalysis generally.

Rieff, Philip, *The Triumph of the Therapeutic* (London: Penguin, 1966).

Sulloway, Frank, *Freud. Biologist of the Mind* (London: Burnett Books, 1979).

Sulloway, Frank, 'Reassessing Freud's case histories: The social construction of psychoanalysis', *Isis* 82 (1991), pp. 245–275.

Yerushalmi, Yosef Hayim, *Freud's Moses. Judaism Terminable and Interminable* (New Haven, CT: Yale University Press, 1991). The best book on the question of Freud's relation to Judaism.

Zaretsky, Eli, *Secrets of the Soul: A Social and Cultural History of Psychoanalysis* (New York: Knopf, 2004).

Index